CW00696650

Lyapunov Matrix Equation in System Stability and Control

Zoran Gajić
Department of Electrical and Computer Engineering
Rutgers University

Muhammad Tahir Javed Qureshi
Hewlett-Packard Corporation

Dover Publications, Inc.
Mineola, New York

Bibliographical Note

This Dover edition, first published in 2008, is an unaltered republication
of the work first published by Academic Press, San Diego, California, in
1995.

Library of Congress Cataloging-in-Publication Data

Gajić, Zoran.
    Lyapunov matrix equation in system stability and control / Zoran
Gajić and Muhammad Tahir Javed Qureshi.—Dover ed.
      p. cm.
    Originally published: San Diego, California : Academic Press, 1995.
    ISBN-13: 978-0-486-46668-2
    ISBN-10: 0-486-46668-X
    1. Control theory. 2. Lyapunov stability. I. Qureshi, Muhammad Tahir
Javed. II. Title.

QA402.3.G3387 2008
003'.85—dc22

                         2007049458

Manufactured in the United States of America
Dover Publications, Inc., 31 East 2nd Street, Mineola, N.Y. 11501

Dedicated to my father Radivoj Gajić
Zoran Gajić

To my wife Shamaila Tahir
Muhammad Qureshi

Contents

Preface

This book is intended for a wide readership including engineers, applied mathematicians, computer scientists, and graduate students who seek a comprehensive view of the main results on the Lyapunov matrix equation. The book presents different techniques for solving and analyzing the algebraic, differential, and difference Lyapunov matrix equations of continuous-time and discrete-time systems. The Lyapunov and Lyapunov-like equations arise in many different prospectives such as control theory, system theory, system identification, linear algebra, optimization, differential equations, boundary value problems and partial differential equations, mechanical engineering, power systems, signal processing, large space flexible structures, communications, and the like. Therefore, its solution is of great interest. The book provides easy and quick references for the solution of many engineering and mathematical problems related to the Lyapunov matrix equations. Because both the mathematical development and the applications are considered, this book is useful for solving problems as well as for research purposes.

In this book we are concerned with the "pure" Lyapunov equation, and only in rare cases the Lyapunov-like equations are discussed. The continuous and discrete Lyapunov matrix equations are considered in three categories: (1) explicit solutions, (2) bounds of the solutions main attributes (such as eigenvalues, determinant, trace), and (3) numerical solutions. The advancements made so far in all these categories are the topics of this book. Different approaches are compared, where possible, in order to demonstrate the efficiency of any particular method. In addition, the recent results on the stability robustness, sensitivity of the Lyapunov equation, parallel algorithms and iterative methods for numerical solution of high dimensional algebraic Lyapunov equations. Also, the Lyapunov matrix equations corresponding to jump parameter linear systems, singularly perturbed and weakly coupled systems are included in this book. Several examples of real-world systems are given throughout of the book in order to demonstrate the effectiveness of the presented methods and algorithms. The book covers research work of more than 250 available journal papers on the Lyapunov matrix equation published in or before December of 1994, and the recent research work by the authors and their coworkers.

The authors are thankful for support and contributions from Professors T-Y. Li, P. Milojević, B. Petrović, and N. Puri, our colleagues Drs. X. Shen and M. Lim, graduate students I. Borno and V. Radisavljević. For technical support, we are indebted to J. Li, I. Seskar, and Dr. A. Kolarov.

Z. Gajić and M. Qureshi
Piscataway, NJ, USA
February 1995

Chapter One

Introduction

The Lyapunov and Lyapunov-like matrix equations appear in many different engineering and mathematical perspectives such as control theory, system theory, optimization, power systems, signal processing, linear algebra, differential equations, boundary value problems, large space flexible structures, and communications, (Dou, 1966; Barnett and Storey, 1970; Kwakernaak and Sivan, 1972; Kreisselmeier, 1972; Balas, 1982; Wonham, 1985; Hodel and Poolla, 1992). It is named after the Russian mathematician Alexander Mikhailovitch Lyapunov (Shcherbakov, 1992; Axelby and Parks, 1992), who in 1892, in his doctoral dissertation, introduced the famous stability theory of linear and nonlinear systems (Lyapunov, 1892). A complete English translation of Lyapunov's 1892 doctoral dissertation is published in *International Journal of Control* in March of 1992. According to his definition of stability, so-called stability in the sense of Lyapunov, one can check the stability of a system by finding some functions, called the Lyapunov functions. There is no general procedure for finding a Lyapunov function for nonlinear systems, but for linear time invariant systems, the procedure comes down to the problem of solving the matrix Lyapunov equation. Since linear systems are mathematically very convenient and give fairly good approximations for nonlinear systems, mathematicians and engineers very often base their analysis on the linearized models. Therefore, the solutions of the Lyapunov matrix equations give insight into the behavior of dynamical systems.

The Lyapunov equation is encountered not only in studying the stability of linear systems, but also in other fields. In mathematics Lyapunov-like equations were subject of research since the beginning of this century (Wedderburn, 1904). The quadratic performance measure of a linear feedback system is given in terms of the solution of the Lyapunov equation. For stochastic linear systems driven by white noise, the solution of the Lyapunov equation represents the variance of the state vector. Many other control and system theory problems are based on the Lyapunov and/or Lyapunov-like equations such as: concepts of controllability and observability grammians (Chen, 1984), balancing transformation (Moore, 1981), stability robustness to parameter variations (Patel and Toda, 1980; Yedavalli, 1985), reduced-order modeling and control (Hyland and Bernstein, 1985, 1986; Bernstein and Hyland, 1985; Safonov and Chiang, 1989), filtering with singular measurement noise (Haddad and Bernstein, 1987; Halevi, 1989), power systems (Ilic, 1989), large space flexible structures (Balas, 1982), estimator design (Chen, 1984). The Lyapunov and/or Lyapunov-like equations also appear in differential games (Petrovic and Gajic, 1988), singular systems (Lewis and Ozcaldiran, 1989), signal processing (Anderson et al., 1986; Agathoklis, 1988; Lu et al., 1992), beam gridworks problem in mechanical engineering (Ma, 1966), differential equations (Dou, 1966), boundary value problems in partial differential equations (Kreisselmeier, 1972), and interpolation problems for rational matrix functions (Lerer and Rodman, 1993).

Due to broad applications, the Lyapunov matrix equation has been subject of very active research for the past thirty years. Although Lyapunov theory was introduced at the end of the nineteenth century, it was not recognized for its vast applications until the 1960s. Since then it has become a major part in controls, system theory, and other fields. Around 1965, some researchers like MacFarlane, Barnett and Storey, Chen and Shieh, Bingulac, and Lancaster presented solutions to the Lyapunov matrix equation. In the 1970s when growing use of digital computers became part of almost every scientific field, the need for efficient numerical solution was felt. This resulted in celebrated algorithms for numerical solution of the continuous-time algebraic Lyapunov equation (Bartels and Stewart, 1972; Golub et al., 1979; Hammarling, 1982). Digital technology in industry also spelled out the need for the solution of the Lyapunov matrix equation for discrete systems, which is slightly different from that for continuous systems. In the years to come, we have seen a lot of results on the bounds of the solution for both continuous-time and discrete-time Lyapunov equations, like trace, determinant, and eigenvalue bounds. Many researchers, especially Mori and Komaroff, contributed a lot to clarify the issue about the best possible bounds for the attributes of the solution of the Lyapunov equations, but the research is still going on. So far we have seen more than 250 journal papers on the Lyapunov equation and related topics.

Lyapunov Equations

This book describes the solutions of the Lyapunov matrix equations and related problems for both continuous-time and discrete-time systems. The general differential Lyapunov matrix equation corresponding to time varying continuous systems is given by

$$A^T(t)P(t) + P(t)A(t) + Q(t) = \dot{P}(t), \quad P(t_0) = P_0 \qquad (1.1)$$

where $P(t)$ is $n \times n$ solution matrix of unknown entries, $A(t)$ is $n \times n$ system matrix, and $Q(t)$ is $n \times n$ symmetric matrix chosen according to computational convenience and in most cases is positive semi-definite. If the system is time invariant then $\dot{P} = 0$ (steady state) so that equation (1.1) becomes the algebraic Lyapunov matrix equation, that is

$$A^T P + PA + Q = 0 \qquad (1.2)$$

The corresponding Lyapunov matrix difference equation for discrete-time varying systems is given by

$$A^T(k)P(k)A(k) + Q(k) = P(k+1), \quad P(k_0) = P_0 \qquad (1.3)$$

and for time invariant discrete systems at the steady state, $(P(k+1) = P(k) = P)$, we have the discrete version of the algebraic Lyapunov matrix equation

$$A^T PA + Q = P \qquad (1.4)$$

The so-called unified continuous-discrete-time algebraic and differential-difference Lyapunov equations have been introduced recently by (Middleton and Goodwin, 1990). Due to lack of research results available in the literature, these equations will not be studied in this book.

Existence Conditions

The solutions of differential and difference Lyapunov equations always exist provided that the matrices A and Q are continuous and bounded in time. However, when the system matrix A is unstable the corresponding differential (difference) Lyapunov equations have infinite escape time.

For time invariant case, the condition for existence of a unique solution of (1.2) is given in many standard texts, for example (Lancaster and Tismenetsky, 1985, page 414). *The necessary and sufficient condition for the existence of a unique solution of (1.2) is that no two eigenvalues of A add up to zero, that is*

$$\lambda_i + \lambda_j \neq 0, \quad i, j = 1, 2, ..., n \qquad (1.5)$$

Justification of this result will be given in Section 2.1.1. Note that condition (1.5) is satisfied if A is an asymptotically stable matrix in the continuous-time domain (all eigenvalues of A are located in the closed left half plane).

An analogous result holds for the discrete algebraic Lyapunov equation. *The condition for the existence of a unique solution of (1.4) is that no two eigenvalues have product equal to one, that is*

$$\lambda_i \lambda_j \neq 1, \quad i, j = 1, 2, ..., n \qquad (1.6)$$

This condition is obviously satisfied if A is asymptotically stable in the discrete-time domain (all eigenvalues of A are strictly inside of a unit circle). *In this book we will assume that A is an asymptotically stable matrix*, and Q is any positive semi-definite matrix, unless otherwise specified.

Although it is possible and worthwhile to describe the development of equations (1.1)-(1.4) in all applications, we will mention in this chapter only the most commonly used ones. Some other applications will be discussed in Chapter 9. In addition, many real world systems whose analysis and design are related to the Lyapunov equations will be presented throughout of the book. For example, we have considered models of F-8 aircraft, steam power system, voltage regulator, synchronous machine connected to an infinite bus, distillation column, catalytic cracker, gas absorber, industrial reactor, L-1011 fighter aircraft, hydroturbine governors, DC motor, magnetic tape control system, one-dimensional diffusion process, aircraft under wind disturbances, and chemical plant.

1.1 Stability of Linear Systems

According to the Lyapunov stability theory (Lyapunov, 1892), the stability of dynamical systems can be determined in terms of certain scalar functions known as Lyapunov functions (Halanay and Rasvan, 1993). This can be done for both linear and nonlinear systems in both continuous-time and discrete-time domains.

Consider a continuous-time invariant linear system

$$\dot{x}(t) = Ax(t), \qquad x(t_0) = x_0 \qquad (1.7)$$

where $x(t) \in \Re^n$ is the system state vector and $A \in \Re^{n \times n}$ is the system matrix. The following stability definitions are very well known (Chen, 1984).

Definition 1.1 The linear system (1.7) is *stable* if all eigenvalues of the matrix A are in the closed left half of the complex plane and those on the imaginary axis are distinct roots of the minimal polynomial of A.

$\triangle$

Definition 1.2 The linear system (1.7) is *asymptotically stable* if all eigenvalues of A are in the open left half of the complex plane.

$\triangle$

Note that Definition 1.1 corresponds to the Lyapunov stability definition, so that "stable" used in this definition also means "stable in the sense of Lyapunov."

The celebrated Lyapunov stability theorem is formulated as follows (for more details see (Khalil, 1992)).

Theorem 1.1 *The equilibrium point* $x = 0$ *of a time invariant dynamical system is stable (in the sense of Lyapunov) if there exists a continuously differentiable scalar function* $V(x)$ *such that along the system trajectories the following is satisfied*

$$V(x) > 0, \qquad V(0) = 0 \tag{1.8a}$$

$$\dot{V}(x) = \frac{dV}{dt} = \frac{\partial V}{\partial x}\frac{dx}{dt} \leq 0 \tag{1.8b}$$

If the condition (1.8b) is a strict inequality then the equilibrium point $x = 0$ *is asymptotically stable.*

$\square$

It is easy to show that for a linear system (1.7) a Lyapunov function can be chosen as a quadratic one, that is

$$V(x) = x^T P x, \qquad P = P^T > 0 \tag{1.9}$$

which with use of (1.7) leads to

$$\dot{V}(x) = x^T\left(A^T P + PA\right)x$$

that is, the system is asymptotically stable if the following condition is satisfied

$$A^T P + PA < 0$$

or

$$A^T P + PA = -Q, \qquad Q = Q^T \tag{1.10}$$

where Q is any positive definite matrix. Now, we are able to formulate the Lyapunov stability theory for linear continuous-time invariant systems.

Theorem 1.2 *The linear time invariant system (1.7) is asymptotically stable if and only if for any* $Q = Q^T > 0$ *there exists a unique* $P = P^T > 0$ *such that (1.10) is satisfied.*

$\square$

Note that Theorem 1.2 can be generalized to include the case $Q = C^T C \geq 0$. *In that case the system (1.7) is asymptotically stable if and only if the pair* (A, C) *is observable and the algebraic Lyapunov equation (1.10) has a unique positive definite solution* (see for example, Chen, 1984, page 414). The observability of the pair (A, C) can be relaxed to its detectability. This is natural since the detectability implies the observability of the modes which are not asymptotically stable.

Dual theorems to Theorems 1.1 and 1.2 can be stated for the stability of discrete-time systems. For a linear discrete-time system

$$x(k + 1) = Ax(k), \qquad x(k_0) = x_0 \qquad (1.11)$$

a Lyapunov function has a quadratic form which must satisfy (Kalman and Bertram, 1960; Ogata, 1987)

$$\begin{aligned} V(k) &= x^T(k)Px(k) > 0 \\ \Delta V(k) &= V(k + 1) - V(k) \le 0 \end{aligned} \qquad (1.12)$$

Since

$$\begin{aligned} V(k + 1) - V(k) &= x^T(k + 1)Px(k + 1) - x^T(k)Px(k) \\ &= x^T(k)\big(A^T PA - P\big)x(k) \le 0 \end{aligned}$$

the stability requirement imposed in (1.12) leads (similarly to the continuous-time argument) to the discrete-time algebraic Lyapunov equation (1.4), which for asymptotic stability, according to the Lyapunov theory (dual result to Theorem 1.2), must have a unique positive definite solution for some positive definite matrix Q.

The general Lyapunov stability theory for time varying systems is more complex than for the time invariant ones (Khalil, 1992). Let us just mention that a linear time varying continuous-time system

$$\dot{x}(t) = A(t)x(t) \qquad (1.13)$$

with a piecewise continuous system matrix $A(t)$ has a Lyapunov function of the form

$$V(x, t) = x^T(t)P(t)x(t) \qquad (1.14)$$

where positive definite matrix $P(t)$ satisfies the Lyapunov differential equation

$$-\dot{P}(t) = A^T(t)P(t) + P(t)A(t) + Q(t) \qquad (1.15)$$

with $Q(t)$ being continuous, symmetric, and positive definite. Similarly, for time varying discrete linear systems

$$x(k+1) = A(k)x(k) \qquad (1.16)$$

a Lyapunov function is given by

$$V(x, k) = x^T(k)P(k)x(k) \qquad (1.17)$$

where $P(k)$ is the positive definite solution of the difference Lyapunov equation

$$P(k) = Q(k) + A^T(k)P(k+1)A(k), \quad Q(k) > 0, \ \forall k \qquad (1.18)$$

1.2 Variance of Linear Stochastic Systems

The linear differential (difference) system driven by white noise is a very convenient model for formulating and solving linear control problems that involve disturbances, uncertainties, and noise. Such a linear differential system is modeled as

$$\dot{x}(t) = A(t)x(t) + \Gamma(t)w(t), \quad E\{x(t_0)\} = \bar{x}_0, \ var(x(t_0)) = P_0 \qquad (1.19)$$

where $x(t)$ is the state vector and $w(t)$ is a zero-mean Gaussian white noise stochastic process with intensity matrix $V(t)$. Assuming that noise is also uncorrelated with the state $x(t)$, the variance $P(t) = E[x(t)x^T(t)]$ of this system satisfies the following Lyapunov matrix differential equation (Kwakernaak and Sivan, 1972)

$$\dot{P}(t) = A(t)P(t) + P(t)A^T(t) + \Gamma(t)V(t)\Gamma^T(t), \quad P(t_0) = P_0 \quad (1.20)$$

The intensity matrix V is symmetric and positive semi-definite so that the matrix $\Gamma V \Gamma^T$ is positive semi-definite.

Similarly, the linear difference equation of a linear discrete stochastic system is given by the state equation

$$x(k+1) = A(k)x(k) + \Gamma(k)w(k), \quad E\{x(k_0)\} = \bar{x}_0, \quad var(x(k_0)) = P_0 \tag{1.21}$$

where $x(k)$ is the state vector and $w(k)$ is a zero-mean Gaussian white noise stochastic process with intensity matrix $V(k)$. Assuming that $w(k)$ is uncorrelated with $x(k)$, the variance matrix $P(k) = E[x(k)x^T(k)]$ of the discrete-time system satisfies the following difference Lyapunov equation

$$P(k+1) = A(k)P(k)A^T(k) + \Gamma(k)V(k)\Gamma^T(k), \quad P(k_0) = P_0 \tag{1.22}$$

This equation is the same as (1.3) with A replaced by A^T, and Q replaced by $\Gamma V \Gamma^T$.

If the above continuous-time and discrete-time linear dynamical systems can reach the steady state, the corresponding state variances are given by the algebraic Lyapunov equations having the forms of (1.2) and (1.4).

1.3 Quadratic Performance Measure

In the system design we often measure the system performance with respect to a quadratic functional. For a continuous-time system

$$\dot{x}(t) = Ax(t), \quad x(t_0) = x_0 \tag{1.23}$$

a quadratic functional is defined by

$$J(t) = \frac{1}{2}x^T(t_f)P_{t_f}x(t_f) + \frac{1}{2}\int_t^{t_f} x^T(\tau)Q(\tau)x(\tau)d\tau \tag{1.24}$$

where Q and P_{t_f} are positive semi-definite matrices. Evaluation of (1.24) along trajectories of (1.23) leads to

$$J(t) = \frac{1}{2}x^T(t)P(t)x(t), \quad t_0 \le t \le t_f \tag{1.25}$$

where $P(t)$ is obtained from the following differential Lyapunov equation (Kwakernaak and Sivan, 1972, page 108)

$$A^T(t)P(t) + P(t)A(t) + Q(t) = -\dot{P}(t), \quad P(t_f) = P_{t_f} \quad (1.26)$$

The minus sign multiplying $\dot{P}$ has no significant effect and it indicates that the integration has to be performed backward in time. Note that if $t_f \to \infty$ and assuming that the system matrix A is time invariant and asymptotically stable, and Q is a constant matrix, the expression (1.25) still holds with $P(t) \to P$, where P satisfies the corresponding algebraic Lyapunov equation (1.2).

The corresponding problem in the discrete-time is described by

$$x(k+1) = Ax(k), \quad x(k_0) = x_0$$

and

$$J(k) = \frac{1}{2}x^T(N)P_N x(N) + \frac{1}{2}\sum_{i=k}^{N-1} x^T(i)Qx(i) \quad (1.27)$$

where N is the final time. Note that

$$J(N) = \frac{1}{2}x^T(N)P(N)x(N), \quad P(N) = P_N \quad (1.28)$$

and

$$J(N-1) = \frac{1}{2}x^T(N-1)P(N-1)x(N-1)$$
$$= \frac{1}{2}x^T(N)P(N)x(N) + \frac{1}{2}x^T(N-1)Q(N-1)x(N-1)$$
$$= \frac{1}{2}x^T(N-1)\left[A^T(N-1)P(N)A(N-1) + Q(N-1)\right]x(N-1)$$

Hence

$$P(N-1) = A^T(N-1)P(N)A(N-1) + Q(N-1)$$

Continuing for $J(N-2), J(N-3), \ldots$, we obtain

$$A^T(k)P(k+1)A(k) + Q(k) = P(k), \quad k < N, \quad P(N) = P_N \quad (1.29)$$

with the corresponding value for the performance criterion given by

$$J(k) = \frac{1}{2} x^T(k) P(k) x(k), \quad k_0 \le k \le N \tag{1.30}$$

Equation (1.29) is dual to equation (1.3). Thus, $J(k)$ can be calculated by solving the difference Lyapunov equation (1.29). Similarly, in the case of $N \to \infty$ with constant matrices A and Q, and A being asymptotically stable, the expression (1.30) holds with the matrix P obtained from the corresponding algebraic discrete-time Lyapunov equation (1.4).

1.4 Book Organization

In this book we are concerned with the "pure" Lyapunov equations and only in rare cases we discuss the Lyapunov-like equations. Thus, the solutions to equations (1.1)-(1.4) and the corresponding analytical and numerical results will be presented. From the available results in the literature, it seems that the Lyapunov equation is still an attractive research area. We will present the solutions in three main categories:

1. Explicit solutions;
2. Approximate solutions characterized by different bounds, like eigenvalue bounds, trace bounds, determinant bounds, solution bounds.
3. Numerical solutions suitable for computer calculation.

Chapter 1 comprises an introduction with a brief historical overview of the Lyapunov equation and the main numerical and analytical results obtained. It also emphasizes the importance of the Lyapunov equation in engineering and science by indicating many areas of applications of this famous equation.

Chapter 2 discusses the *continuous-time algebraic Lyapunov equation*. In this chapter we first present the explicit solutions of the algebraic Lyapunov equations corresponding to continuous-time systems. These solutions are in fact the classic methods based on the conversion of the algebraic Lyapunov equation into a system of linear equations (MacFarlane, 1963; Bingulac, 1970). In addition, some series expansion methods (Mori, et al., 1986a) and the skew-symmetric matrix approach (Barnett and Storey, 1966a, 1966b; Barnett, 1974) are also included in the category of explicit solutions of the algebraic Lyapunov equation. Next, we study the bounds of the solution of the algebraic Lyapunov equation. In that context the eigenvalue bounds, trace bounds, and determinant bounds are considered. The results presented give the complete review and summary of all results available in the literature with emphasis on the work of (Yasuda and Hirai, 1979; Karanam, 1981, 1982; Mori et al., 1986a; Wang et al., 1986; Komaroff, 1988, 1990, 1992). It is important to point out that no final word is given about the best possible bounds so that further research is needed in that direction. In the last part of this chapter, the numerical methods for solving the algebraic Lyapunov equation are presented. We first review some classic and interesting methods due to (Davison and Man, 1968; Smith, 1968; Hoskins et al., 1977) and then present the most important one, that is, the algorithm of (Bartels and Stewart, 1972) and comment on its relation to the similar algorithms of (Golub et. al., 1979) and (Hammarling, 1982).

Chapter 3 discusses solutions and properties of the *discrete-time algebraic Lyapunov equation*. Similarly to the previous chapter, here the discrete algebraic Lyapunov equation is considered in three categories, namely, explicit solutions, solution bounds, and numerical solutions. In the context of explicit solutions we consider the bilinear transformation technique (Popov, 1964; Power, 1967, 1969), skew-symmetric matrix approach (Barnett, 1974), and Jordan form technique (Heinen, 1972). Several results are presented for the solution bounds. According to the work of Mori and his coworkers, and Komaroff, the leading researchers in this field, no definite answers have been obtained yet about the best possible bounds for the solution attributes (eigenvalues, trace, determinant) of the discrete-time algebraic Lyapunov equation. Obtained bounds are mostly based on different mathematical inequalities for which it is hard to make definite analytical comparisons. As far as the numerical solutions are concerned, we present the results of (Barraud, 1977), which resemble the Bartels-Stewart algorithm for numerical solution of the continuous algebraic Lyapunov equation.

In Chapter 4, the explicit solutions, bounds of the solution's main attributes (eigenvalues, trace, determinant), and numerical solutions of the *differential and difference Lyapunov equations* are presented. The explicit solutions have been presented for diagonalizable matrix A only. The bounds considered are based on a few results available in the literature (Mori et al., 1986b, 1987; Hmamed, 1990). We also present the solution bounds for the differential Lyapunov equation according to the work of (Geromel and Bernussou, 1979). Moreover, the obtained solution bounds at the steady state give the bounds for the solution of the continuous algebraic Lyapunov equation. Numerical solutions present mostly work of (Subrahmanyam, 1986) for differential time varying Lyapunov equation, and the classic work of (Davison, 1975) applicable to the corresponding time invariant equation. In addition, we consider the differential and difference Lyapunov equations with small parameters corresponding to singularly perturbed and weakly coupled systems (Gajic and Shen, 1989, 1993), where the order-reduction is achieved by using suitable state transformations so that the original (global) Lyapunov equations are decomposed into the reduced-order subsystem local Lyapunov equations. At the end of this chapter, the coupled differential Lyapunov equations are discussed.

Chapter 5 presents solutions for the *algebraic Lyapunov equations with small parameters* (Gajic et al., 1989; Gajic et al., 1990; Gajic and Shen, 1993) corresponding to singularly perturbed and weakly coupled systems. The recursive algorithms for the solution of these equations are derived in the most general case when the problem matrices are functions of a small perturbation parameter. The numerical decompositions have been achieved, so that only low-order systems are involved in algebraic computations. The introduced recursive methods can be implemented as synchronous parallel algorithms. Both continuous-time and discrete-time versions of the algebraic Lyapunov equations are studied. It is shown that the singular perturbation recursive methods converge with the rate of convergence of $O(\epsilon)$, whereas the recursive methods for weakly coupled linear systems converge faster, that is, with the rate of convergence of $O(\epsilon^2)$, where ϵ is a small perturbation parameter.

Chapter 6 studies *stability robustness of linear systems and sensitivity of the algebraic Lyapunov equations*. The stability robustness presentation is done in detail, and it is mostly based on the work of (Patel and Toda, 1980; Yedavalli, 1985; Zhou and Khargonekar, 1987). Sensitivity of the continuous-time algebraic Lyapunov equation is considered according to (Jonckheere, 1984; Hewer and Kenney, 1988), whereas the corresponding discrete-time sensitivity results are presented by following the work of (Gahinet et al., 1990). Only the main sensitivity results are presented in this chapter. For deeper study of this attribute of the Lyapunov equation the reader is referred to the very comprehensive papers of (Hewer and Kenney, 1988; Gahinet et al., 1990).

In Chapter 7, the *parallel algorithms* (Hodel and Poolla, 1992) and the *iterative methods* for solving large scale Lyapunov equations are presented (Smith, 1968; Wachspress, 1988, 1992; Lu and Wachspress, 1991; Starke, 1991, 1992). These important methods are modern trend in numerical analysis of the algebraic Lyapunov equation. In addition, parallel algorithms for solving the coupled algebraic Lyapunov equations of jump linear systems are considered for both continuous-time and discrete-time domains by following the work of (Borno, 1995; Borno and Gajic, 1995).

Chapter 8 deals with the *Lyapunov iterations* for solving nonlinear algebraic equations arising in the control theory. We consider here the Lyapunov iterations technique in the context of the linear-quadratic optimal control problem (Kleinman 1968), output feedback control (Moerder and Calise, 1985), applications to the differential games (Petrovic and Gajic, 1988), jump linear systems (Gajic and Borno, 1995), and the solution of the general coupled algebraic Riccati equations (Gajic and Li, 1988; Li and Gajic, 1994).

In Chapter 9, we give a brief review of the *Sylvester equation* (Lyapunov-like equation), discuss different applications of the Lyapunov equation and comments on related topics. At the end, in Appendix, some matrix inequalities appearing in the book are summarized.

Since this book may be used by the graduate students and researchers as a quick reference to the highlights on the Lyapunov equation, the presentation is such that every chapter is as much as independent as possible so that the reader may go directly to the desired chapter or section by following either the table of contents or the book's index. On the other hand, the authors' intention is to provide a comprehensive treatment of all important aspects of the Lyapunov equation so that the book precisely and concisely considers all main results of this already application broad and research challenging area.

1.5 References

1. Agathoklis, P., "The Lyapunov equation for n-dimensional discrete systems," *IEEE Trans. Automatic Control*, vol.35, 448–451, 1988.

2. Anderson, B., P. Agathoklis, E. Jury, and M. Mansour, "Stability and the matrix Lyapunov equation for discrete 2–dimensional systems," *IEEE Trans. Circuits and Systems*, vol.33, 261–266, 1986.

3. Axelby, G. and P. Parks, "Lyapunov Centenary," *Automatica*, vol.28, 863–864, 1992.

4. Balas, M., "Trends in large space structure control theory: Fondest hopes, wildest dreams," *IEEE Trans. Automatic Control*, vol.27, 522–535, 1982.

5. Barnett, S., "Simplification of Lyapunov matrix equation $A^T P A - P = -Q$," *IEEE Trans. Automatic Control*, vol.19, 446-447, 1974.

6. Barnett, S. and C. Storey, "Stability analysis of constant linear systems by Liapunov second method," *Electronics Letters*, vol.2, 165–166, 1966a.

7. Barnett, S. and C. Storey, "Solution of the Liapunov matrix equation," *Electronics Letters*, vol.2, 466–467, 1966b.

8. Barnett, S. and C. Storey, *Matrix Methods in Stability Theory*, Nelson, London, 1970.

9. Barraud, A., "A numerical algorithm to solve $A^T X A - X = Q$," *IEEE Trans. Automatic Control*, vol.22, 883–885, 1977.

10. Bartels, R. and G. Stewart, "Algorithm 432, solution of the matrix equation $AX + XB = C$," *Comm. Ass. Computer Machinery*, vol.15, 820–826, 1972.

11. Bernstein, D. and D. Hyland, "The optimal projection equation for reduced-order state estimation," *IEEE Trans. Automatic Control*, vol.30, 583–585, 1985.

12. Bingulac, S., "An alternate approach to expanding $PA + A^T P = -Q$," *IEEE Trans. Automatic Control*, vol.15, 135-136, 1970.

13. Borno, I., *Parallel Algorithms for Optimal Control of Linear Jump Parameter Systems and Markov Processes*, Doctoral Dissertation, Rutgers University, 1995.

14. Borno, I. and Z. Gajic, "Parallel algorithm for solving coupled algebraic Lyapunov equations of discrete-time jump linear systems," *Computers & Mathematics with Appl.*, vol.29, in press, 1995.

15. Chen, C., *Linear System Theory and Design*, Holt, Rinehart and Winston, New York, 1984.

16. Davison, E. "The numerical solution of $X = A_1 X + X A_2 + D$, $X(0) = C$," *IEEE Trans. Automatic Control*, vol.20, 566–567, 1975.

17. Davison, E. and F. Man, "The numerical solution of $A^T Q + QA = -C$," *IEEE Trans. Automatic Control*, vol.13, 448-449, 1968.

18. Dou, A., "Method of undetermined coefficients in linear differential systems and the matrix equation $YB - AY = F$," *SIAM J. Appl. Math.*, vol.14, 691–696, 1966.

19. Gajic, Z. and I. Borno, "Lyapunov iterations for optimal control of jump linear systems at steady state," *IEEE Trans. Automatic Control*, to appear, 1995.

20. Gajic, Z. and T. Li, "Simulation results for two new algorithms for solving coupled algebraic Riccati equations," *Third Int. Symposium on Differential Games and Applications*, Sophia Antipolis, France, 1988.

21. Gajic, Z., D. Petkovski, and N. Harkara, "The recursive algorithm for the optimal static output feedback control problem of linear singularly perturbed systems," *IEEE Trans. Automatic Control*, vol.34, 465-468, 1989.

22. Gajic, Z., D. Petkovski, and X. Shen, *Singularly Perturbed and Weakly Coupled Linear Control Systems: A Recursive Approach*, Springer Verlag, Berlin, 1990.

23. Gajic, Z. and X. Shen, "Decoupling transformation for weakly coupled linear systems," *Int. J. Control*, vol.50, 1517-1523, 1989.

24. Gajic, Z. and X. Shen, *Parallel Algorithms for Optimal Control of Linear Large Scale Systems*, Springer Verlag, London, 1993.

25. Gahinet, P., A. Laub, C. Kenney, and G. Hewer, "Sensitivity of the stable discrete-time Lyapunov equation," *IEEE Trans. Automatic Control*, vol.35, 1209–1217, 1990.

26. Geromel, J. and J. Bernussou, "On bounds of Lyapunov matrix equation," *IEEE Trans. Automatic Control*, vol.24, 482-483, 1979.

27. Golub, G., S. Nash, and C. Loan, "A Hessenberg-Schur method for the problem $AX + XB = C$," *IEEE Trans. Automatic Control*, vol.24, 909-913, 1979.

28. Haddad, W. and D. Bernstein, "The optimal projection equations for reduced-order state estimation: the singular measurement noise," *IEEE Trans. Automatic Control*, vol.32, 1135-1143, 1987.

29. Halanay A. and V. Rasvan, *Applications of Lyapunov Methods in Stability*, Kluwer, Dordrecht, The Netherlands, 1993.

30. Halevi, Y., "The optimal reduced-order estimator for systems with singular measurement noise," *IEEE Trans. Automatic Control*, vol.34, 777–781, 1989.

31. Hammarling, S., "Numerical solution of the stable, non-negative definite Lyapunov equation," *IMA J. Numerical Analysis*, vol.2, 303-323, 1982.

32. Heinen, J., "A technique for solving the extended discrete Lyapunov matrix equation," *IEEE Trans. Automatic Control*, vol.17, 156-157, 1972.

33. Hewer, G. and C. Kenney, "The sensitivity of the Lyapunov equation," *SIAM J. Control and Optimization*, vol.26, 321-344, 1988.

34. Hmamed, A., "Differential and difference Lyapunov equations: simultaneous eigenvalue bounds," *Int. J. Systems Science*, vol.21, 1335-1344, 1990.

35. Hodel, A. and K. Poolla, "Parallel solution of large Lyapunov equations," *SIAM J. Matrix Anal. Appl.*, vol.13, 1189–1203, 1992.

36. Hoskins, W., D. Meek, and D. Walton, "The numerical solution of $A^T Q + QA = -C$," *IEEE Trans. Automatic Control*, vol.22, 882-885, 1977.

37. Hyland, D. and D. Bernstein, "The optimal projection equations for model reduction and the relationships among the methods of Wilson, Skelton, and Moore," *IEEE Trans. Automatic Control*, vol.30, 1201–1211, 1985.

38. Hyland, D. and D. Bernstein, "The optimal projection equations for finite dimensional fixed-order dynamic compensation of infinite-dimensional systems," *SIAM J. Control and Optimization*, vol.24, 122–151, 1986.

39. Ilic, M., "New approaches to voltage monitoring and control," *IEEE Control System Magazine*, vol.9, 3–11, 1989.

40. Jonckheere, E., "New bounds on the sensitivity of the solution of the Lyapunov equation," *Linear Algebra and Its Appl.*, vol.60, 57–64, 1984.

41. Kalman, R and J. Bertram, "Control system analysis and design via the second method of Lyapunov: II Discrete time systems, *Trans. ASME J. Basic Eng., Series D.*, 394–400, 1960.

42. Karanam, V., "Lower bounds on the solution of Lyapunov matrix and algebraic Riccati equation," *IEEE Trans. Automatic Control*, vol.26, 1288-1290, 1981.

43. Karanam, V., "Eigenvalue bounds for algebraic Riccati and Lyapunov equations," *IEEE Trans. Automatic Control*, vol.27, 461-463, 1982.

44. Khalil, H., *Nonlinear Systems*, Macmillan, New York, 1992.

45. Kleinman, D., "On an iterative technique for Riccati equation computations," *IEEE Trans. Automatic Control*, vol.13, 114–115, 1968.

46. Komaroff, N., "Simultaneous eigenvalue lower bounds for Lyapunov matrix equation," *IEEE Trans. Automatic Control*, vol.33, 126-128, 1988.

47. Komaroff, N., "Upper bounds for the eigenvalues of the solution of the Lyapunov matrix equation," *IEEE Trans. Automatic Control*, vol.35, 737-739, 1990.

48. Komaroff, N., "Upper summation and product bounds for solution eigenvalues of the Lyapunov matrix equation," *IEEE Trans. Automatic Control*, vol.37, 1040–1042, 1992.

49. Kreisselmeier, G., "A solution of the bilinear matrix equation $AY + YB = -Q$," *SIAM J. Appl. Math.*, vol.23, 334–338, 1972.

50. Kwakernaak, H. and R. Sivan, *Linear Optimal Control Systems*, Wiley, 1972.

51. Lancaster, P. and M. Tismenetsky, *Theory of Matrices*, Academic Press, New York, 1985.

52. Lerer, L. and L. Rodman, "Sylvester and Lyapunov equations and some interpolation problems for rational matrix functions," *Linear Algebra and Its Appl.*, vol.185, 83–117, 1993.

53. Lewis, F. and K. Ozcaldiran, "Geometric structure and feedback in singular systems," *IEEE Trans. Automatic Control*, vol.34, 450–455, 1989.

54. Li, T. and Z. Gajic, "Lyapunov iterations for solving coupled algebraic Riccati equations of Nash differential games and algebraic Riccati equations of zero-sum games," *Proc. Symp. on Dynamic Games and Appl.*, 489–494, St-Jovite, Canada, 1994.

55. Lu, A. and E. Wachspress, "Solution of Lyapunov equations by alternating direction implicit iteration," *Computers & Mathematics with Appl.*, vol.21, 43–58, 1991.

56. Lu, W., H. Wang, and A. Antoniou, "An efficient method for evaluation of the controllability and observability gramians of 2–D digital filters and systems," *IEEE Trans. on Circuits and Systems*, vol.39, 695–704, 1992.

57. Lyapunov, M., "General stability problem of motion," in Russian, Gostehizdat, Moskow, 1950, third edition of the original Lyapunov paper published in 1892.

58. Ma, E., "A finite series solution of the matrix equation $AX - XB = C$," *SIAM J. Appl. Math.*, vol.14, 490–495, 1966.

59. MacFarlane, A., "The calculation of functionals of the time and frequency response of a linear constant coefficient dynamical system," *Quart. Journal Mech. and Applied Math.*, Pt.2, vol.16, 259–271, 1963.

60. Middleton, R. and G. Goodwin, *Digital Control and Estimation: A Unified Approach*, Prentice Hall, Englewood Cliffs, 1990.

61. Moerder, D. and A. Calise, "Convergence of a numerical algorithm for calculating optimal output feedback gains," *IEEE Trans. Automatic Control*, vol.30, 900-903, 1985.

62. Moore, B., "Principal component analysis in linear systems: Controllability, observability and model reduction," *IEEE Trans. Automatic Control*, vol.26, 17–32, 1981.

63. Mori, T., N. Fukuma, and M. Kuwahara, "Explicit solution and eigenvalue bounds in the Lyapunov matrix equation," *IEEE Trans. Automatic Control*, vol.31, 656-658, 1986a.

64. Mori, T., N. Fukuma, and M. Kuwahara, "On the Lyapunov matrix differential equation," *IEEE Trans. Automatic Control*, vol.31, 868-869, 1986b.

65. Mori, T., N. Fukuma, and M. Kuwahara, "Bounds in the Lyapunov matrix differential equations," *IEEE Trans. Automatic Control*, vol.32, 55-57, 1987.

66. Ogata, K., *Discrete-Time Control Systems*, Prentice Hall, Englewood Cliffs, 1987.

67. Patel, R. and M. Toda, "Quantitative measures of robustness for multivariable systems," *Proc. Joint American Control Conf.*, paper TP8–A, San Francisco, 1980.

68. Petrovic, B. and Z. Gajic, "The recursive solution of linear quadratic Nash games for weakly interconnected systems," *J. Optimization Theory and Appl.*, vol.56, 463-477, 1988.

69. Popov, V., "Hyperstability and optimality of automatic systems with several control functions," *Rev. Roum. Sci. Tech.*, vol.9, 629-690, 1964.

70. Power, H., "Equivalence of Lyapunov matrix equations for continuous and discrete systems," *Electronics Letters*, vol.3, 83, 1967.

71. Power, H., "A note on matrix equation $A^T LA - L = -K$," *IEEE Trans. Automatic Control*, vol.14, 411-412, 1969.

72. Safonov, M. and R. Chiang, "A Schur method for balanced-truncation model reduction," *IEEE Trans. Automatic Control*, vol.34, 729–733, 1989.

73. Shcherbakov, P., "Alexander Mikhailovitch Lyapunov: On the centenary of his doctoral dissertation on stability of motion," *Automatica*, vol.28, 865–871, 1992.

74. Smith, R., "Matrix equation $XA + BX = C$," *SIAM J. Appl. Math.*, vol.16, 198-201, 1968.

75. Starke, G., "SOR for $AX - XB = C$," *Linear Algebra and Its Appl.*, vol.154–156, 355–375, 1991.

76. Starke, G., "SOR-like methods for Lyapunov matrix equations," in *Iterative Methods in Linear Algebra*, 233–240, R. Beauwens and P. deGroen, Eds., North-Holland, Amsterdam, 1992.

77. Subrahmanyam, M., "On a numerical method of solving the Lyapunov and Sylvester equations," *Int. J. Control*, vol.43, 433-439, 1986.

78. Wachspress, E. "Iterative solution of the Lyapunov matrix equation," *Applied Math. Letters*, vol.1, 87–90, 1988.

79. Wachspress, E., "ADI Iterative solution of Lyapunov equations," in *Iterative Methods in Linear Algebra*, 229–231, R. Beauwens and P. deGroen, Eds., North-Holland, Amsterdam, 1992.

80. Wang, S., T. Kuo, and C. Hsu, "Trace bounds on the solution of algebraic matrix Riccati and Lyapunov equations," *IEEE Trans. Automatic Control*, vol.31, 654-656, 1986.

81. Wedderburn, J., "Note on the linear matrix equation," *Proc. Edinburgh Math. Soc.* vol.22, 49–53, 1904.

82. Wonham, W., *Linear Multivariable Control: A Geometric Approach*, Springer, New York, 1985.

83. Yasuda, K. and K. Hirai, "Upper and lower bounds on solution of algebraic Riccati equation," *IEEE Trans. Automatic Control*, vol.24, 483-487, 1979.

84. Yedavalli, R., "Improved measures of stability robustness of linear state space models," *IEEE Trans. Automatic Control*, vol.30, 577-579, 1985.

85. Zhou, K. and P. Khargonekar, "Stability robustness for linear state space models with structured uncertainty," *IEEE Trans. Automatic Control*, vol.32, 621-623, 1987.

Chapter Two

Continuous Algebraic Lyapunov Equation

In this chapter we first present the methods for explicit solution of the algebraic Lyapunov equation corresponding to continuous-time systems. These classic methods are based on the conversion of the algebraic Lyapunov equation into a system of linear equations (MacFarlane, 1963; Bingulac, 1970). The explicit solutions are particularly important when the system matrix has any of the special forms like phase variable canonical form (Barnett and Storey, 1967a; Guidorzi, 1972; Sreeram and Agathoklis, 1991; Xiao et al., 1992) or Schwarz form (Kalman and Bertram, 1960; Ziedan, 1972) since in those cases the obtained results are quite simple. In addition, some series expansion methods (Mori et al., 1986) and the skew-symmetric matrix approach (Barnett and Storey, 1966a, 1966b; Barnett, 1974) are also included in the category of explicit solutions of the algebraic Lyapunov equation.

In the following we study the bounds of the solution of the algebraic Lyapunov equation. In that context we consider the eigenvalue bounds, trace bounds, and determinant bounds. The presentation gives the complete review and summary of all obtained results with emphasis on the work of (Yasuda and Hirai, 1979; Karanam, 1981, 1982; Mori et al., 1986a; Wang et al., 1986; Komaroff, 1988, 1990, 1992). It is important to point out that no final word is given about the best possible bounds so that further research is needed in that direction.

Several short proofs of stated theorems are presented in this chapter together with the fundamental inequalities needed to complete the proofs. For the proofs of other important theorems the reader is referred to the original papers. A summary of all major inequalities used in this book is given in Appendix.

In the last part of this chapter, the numerical methods for solving the algebraic Lyapunov equation are presented. We first review some classic and interesting methods due to (Davison and Man, 1968; Smith, 1968; Hoskins et al., 1977) and then present the most important one, that is, the algorithm of (Bartels and Stewart, 1972) and comment on its relation to the similar algorithms of (Golub et al., 1979) and (Hammarling, 1982). Iterative methods for numerical solution of large scale Lyapunov algebraic equations and the parallel algorithms for solving these equations will be studied in Chapter 7.

2.1 Explicit Solutions

Researchers interested in the Lyapunov equation in the late 1960's and early 1970's spent most of their efforts in finding direct explicit methods for efficient solution of the algebraic Lyapunov matrix equation

$$A^T P + PA + Q = 0 \tag{2.1}$$

where $A, Q, P \in \Re^{n \times n}$. Several techniques were proposed. We group them into expansion methods, skew-symmetric matrix method, and transformation techniques. The transformation techniques are particularly important for the special cases when the system matrix A is in one of the canonical forms.

2.1.1 Expansion Methods

The obvious way to solve (2.1) is to expand it into a system of linear algebraic equations and then use the conventional techniques to solve the obtained linear system. Many of the early approaches used this technique. Since much literature is available for solving simultaneous linear equations, therefore, in this section emphasis will be given on the expansion procedures of the Lyapunov matrix equation into a set of linear algebraic equations. Along with the general structure of the coefficient matrix A, we will explore some special cases, in which matrix A has some special canonical forms, and thus enables us to find the solution matrix P more directly and easily.

In developing the expansion procedures, two main goals have to be kept in mind, namely, 1) simplicity, and 2) number of linear equations which one has to solve. In other words, the procedure should be simple enough to convert the matrix equation into a set of linear equations, and it should yield minimum possible number of equations.

In (Bellman, 1959) a suggestion was given to use the Kronecker product to solve (2.1). The general method for solving the algebraic Lyapunov equation based on the Kronecker product is described in (Barnett and Storey, 1967b). The method is straightforward and simple, but it yields to n^2 linear equations for the n^2 unknown entries of P. The method converts (2.1) into an equivalent system of linear equations as

$$\mathcal{A}x = -b \qquad (2.2)$$

where

$$\mathcal{A} = I \otimes A^T + A^T \otimes I \qquad (2.3)$$

and

$$b^T = [q_{11}, q_{12}, ..., q_{1n}, q_{21}, ..., q_{2n}, ..., q_{n-1,1}, ..., q_{n-1,n}, q_{n1}, ..., q_{nn}]$$

$$x^T = [p_{11}, p_{12}, ..., p_{1n}, p_{21}, ..., p_{2n}, ..., p_{n-1,1}, ..., p_{n-1,n}, p_{n1}, ..., p_{nn}]$$

The symbol $\otimes$ denotes the Kronecker product, defined as

$$A \otimes B = \begin{bmatrix} a_{11}B & a_{12}B & \dots & a_{1n}B \\ a_{21}B & a_{22}B & \dots & a_{2n}B \\ \vdots & \vdots & \ddots & \vdots \\ a_{n1}B & a_{n2}B & \dots & a_{nn}B \end{bmatrix} \in \Re^{nm \times nm} \qquad (2.4)$$

where $A \in \Re^{n \times n}$, $B \in \Re^{m \times m}$. The theory of Kronecker product can be found in many texts on linear algebra, e.g. (Lancaster and

Tismenetsky, 1985).   Obviously, *the unique solution of the obtained algebraic equation (2.2) exists if and only if the matrix $\mathcal{A}$ is nonsingular.* The invertibility of the matrix $\mathcal{A}$ can be checked by examining the eigenvalues of the system matrix A.   It is known (Lancaster, 1970; Lancaster and Tismenetsky, 1985) that if λ denotes the eigenvalues of A, then $\lambda_k(\mathcal{A}) = \lambda_i + \lambda_j$, $i, j = 1, 2, ..., n$; $k = 1, 2, ..., n^2$.   Thus, the existence and uniqueness condition for (2.2), which has already been stated in (1.5), is $\lambda_i + \lambda_j \neq 0, \forall i, \forall j$. This condition is obviously satisfied if A is asymptotically stable, that is, if all the eigenvalues of matrix A lie in the left half of the complex plane.

The Kronecker product method is very simple and elegant, but its use is restricted to the cases when n is small. For large n, the difficulty in solving n^2 linear equations makes it impractical, and therefore, other methods are sought to reduce the number of equations.   Note that this method is also used to solve a more general problem studied in (Lancaster, 1970), where the following algebraic equation was considered

$$\sum_{j=1}^{k} A_j X B_j = C \tag{2.5}$$

with A_j, B_j, and C being known matrices not necessarily square. In (Lu, 1971) the Kronecker product method is used to solve the Lyapunov-like equation

$$A_1 P + P A_2 = C \tag{2.6}$$

also known as *the Sylvester algebraic equation*, with $P, C \in \Re^{m \times n}$, $A_1 \in \Re^{m \times m}, A_2 \in \Re^{n \times n}$.   The theoretical aspects of the algebraic Sylvester equation can be found in (Gantmacher, 1959). In this book we will give a brief review of the Sylvester equation in Chapter 9.

In order to reduce the number of linear equations the symmetry of the matrix P can be exploited.   It can be readily seen that due to the assumption that the matrix Q is symmetric, the solution matrix P is also symmetric. Hence there are only $0.5n(n + 1)$ unknown values, which can be found through $0.5n(n + 1)$ simultaneous linear equations. The first such a procedure is given in (MacFarlane, 1963), where (2.1) is converted into the following equivalent set of equations

$$\mathbf{A}p = -\frac{1}{2}q \tag{2.7}$$

where $\mathbf{A}$ is $0.5n(n+1) \times 0.5n(n+1)$ matrix, p and q are $0.5n(n+1)$ column vectors given by

$$p = [p_{11}, p_{12}, ..., p_{1n}, p_{22}, p_{23}, ..., p_{2n}, ..., p_{n-1,n-1}, p_{n-1,n}, p_{nn}]^T$$

$$q = [q_{11}, 2q_{12}, ..., 2q_{1n}, q_{22}, 2q_{23}, ..., 2q_{2n}, ...,$$
$$q_{n-1,n-1}, 2q_{n-1,n}, q_{nn}]^T$$

An elegant procedure for constructing the matrix $\mathbf{A}$ by using three "if" statements for the method of MacFarlane is given in (Chen and Shieh, 1968). It is outlined below in the form of an algorithm.

Algorithm 2.1:

Construct a table whose row index is (k, l) and column index is (i, j) with the following restrictions:

1. if $k = i, l \neq j \longrightarrow A(l,j)$
    if $k \neq i, l = j \longrightarrow A(k,i)$

2. if $k \neq i, l \neq j$, then
    $k = j, l \neq i \longrightarrow A(l,i)$
    $k \neq j, l = i \longrightarrow A(k,j)$
    $k \neq j, l \neq i \longrightarrow 0$

3. if $k = i, l = j$, then
    $k = j, l = i \longrightarrow A(k,i)$
    $k \neq j, l \neq i \longrightarrow A(k,i) + A(l,j)$

$$\Delta$$

Example 2.1: In Table 2.1 we present an example for constructing the augmented matrix for a system of algebraic equations (2.7) corresponding to the algebraic Lyapunov equation (2.1) with the matrix A given by

$$A = \begin{bmatrix} a_{11} & a_{12} & a_{13} \\ a_{21} & a_{22} & a_{23} \\ a_{31} & a_{32} & a_{33} \end{bmatrix}$$

The entries of the table given below are the corresponding entries of the matrix **A**.

	(k,l)					
(j,i)	(1,1)	(1,2)	(1,3)	(2,2)	(2,3)	(3,3)
(1,1)	a_{11}	a_{21}	a_{31}	0	0	0
(1,2)	a_{12}	$a_{11} + a_{22}$	a_{32}	a_{21}	a_{31}	0
(1,3)	a_{13}	a_{23}	$a_{11} + a_{33}$	0	a_{21}	a_{31}
(2,2)	0	a_{12}	0	a_{22}	a_{32}	0
(2,3)	0	a_{13}	a_{12}	a_{23}	$a_{22} + a_{33}$	a_{32}
(3,3)	0	0	a_{13}	0	a_{23}	a_{33}

Table 2.1: Augmented matrix **A**

Δ

The procedure for obtaining the matrix **A** in (2.7) is not unique. An alternate approach for such an expansion, given in (Bingulac, 1970), is more suitable for computer calculations. This method converts (2.1) into the following form

$$\mathbf{U}p = -r \qquad (2.8)$$

where p is the column vector as before, but r is slightly different

$$r^T = [q_{11}, q_{12}, ..., q_{1n}, q_{22}, q_{23}, ..., q_{2n}, ..., q_{n-1,n-1}, q_{n-1,n}, q_{nn}]$$

The algorithm used to compute **U** requires an auxiliary $n \times n$ matrix L with integer entries. Construction of matrix **U** is done through the following steps (Bingulac, 1970).

Algorithm 2.2:

Step 1. Construct the $n \times n$ matrix L, given by

$$L = \begin{bmatrix} 1 & 2 & 3 & \ldots & \ldots & n \\ 2 & n+1 & n+2 & \ldots & \ldots & 2n-1 \\ 3 & n+2 & 2n & 2n+1 & \ldots & 3n-3 \\ 4 & n+3 & 2n+1 & 3n-2 & \ldots & \ldots \\ \vdots & \vdots & \vdots & \vdots & \ddots & \vdots \\ \vdots & \vdots & \vdots & \vdots & N-2 & N-1 \\ n & 2n-1 & 3n-3 & \ldots & N-1 & N \end{bmatrix}$$

where $N = n(n+1)/2$.

Step 2. Construct the $N \times N$ matrix $V = \{v_{xy}\}$ with

$$v_{xy} = a_{ji}$$

where the indices x and y $(x, y = 1, 2, \ldots, N)$ are given by the following elements of the auxiliary matrix L:

$$x = L_{ik}, \quad y = L_{jk}, \quad i, j, k = 1, 2, \ldots, n$$

Step 3. The required matrix U is obtained by multiplying by 2 all the elements of V whose row indices correspond to diagonal elements of the matrix L, that is, $1, n+1, 2n, 3n-2, \ldots, N-2$, and N.

Δ

Example 2.2: Consider the case $n = 3$. Then, L is equal to

$$L = \begin{bmatrix} 1 & 2 & 3 \\ 2 & 4 & 5 \\ 3 & 5 & 6 \end{bmatrix}$$

and the matrix V is

$$V = \begin{bmatrix} a_{11} & a_{21} & a_{31} & 0 & 0 & 0 \\ a_{12} & a_{11}+a_{22} & a_{32} & a_{21} & a_{31} & 0 \\ a_{13} & a_{23} & a_{11}+a_{33} & 0 & a_{21} & a_{31} \\ 0 & a_{12} & 0 & a_{22} & a_{32} & 0 \\ 0 & a_{13} & a_{12} & a_{23} & a_{22}+a_{33} & a_{32} \\ 0 & 0 & a_{13} & 0 & a_{23} & a_{33} \end{bmatrix}$$

which is the same as in Table 2.1. Finally, we obtain $\mathbf{U}$ by multiplying row number 1, 4, and 6 of the matrix V by 2.

$\triangle$

In iterative calculations, where the solution of (2.8) and construction of the matrix $\mathbf{U}$ are in an iterative loop, this method is more suitable from computer time point of view than the method presented in Algorithm 2.1. FORTRAN code which implements this algorithm is given in (Bingulac, 1970).

A method for finding the solution of the algebraic Lyapunov equation that involves calculation of the matrix exponential $e^{A^T t}$ is described in (Mori et al., 1986). The need for evaluating such an exponential is motivated by the fact that (2.1) has the solution in the form

$$P = \int_0^\infty e^{A^T t} Q e^{At} dt \tag{2.9}$$

Using the Cayley-Hamilton theorem the matrix exponential $e^{A^T t}$ can be expressed as

$$e^{A^T t} = a_1(t)I_n + a_2(t)A^T + \ldots + a_n(t)\left(A^T\right)^{n-1} \tag{2.10}$$

where $a_1(t), a_2(t), \cdots, a_n(t)$ are calculated by conventional methods. Define $Q = \Gamma\Gamma^T, \Gamma \in \Re^{n \times r}$ and

$$M = \left[\Gamma, A^T\Gamma, \left(A^T\right)^2\Gamma, \ldots, \left(A^T\right)^{n-1}\Gamma\right] \tag{2.11}$$

The solution of (2.1) as given by (2.9) can be obtained by substituting the expansion for the matrix exponential (2.10) into (2.9), which leads to

$$P = MHM^T \tag{2.12}$$

where

$$H = G \otimes I_r \in \Re^{nr \times nr}, \quad G = G^T = \{g_{ij}\} \in \Re^{n \times n}$$

$$g_{ij} = \int_0^\infty a_i(t)a_j(t)dt \tag{2.13}$$

The symbol $\otimes$ denotes the Kronecker product already defined in (2.4).

Example 2.3: Consider the following matrices

$$A = \begin{bmatrix} 0 & 1 \\ -2 & -3 \end{bmatrix}, \quad \Gamma = \begin{bmatrix} 1 \\ 0 \end{bmatrix}$$

The eigenvalues of A are $\lambda_1 = -1$, $\lambda_2 = -2$. The coefficients $a_1(t)$ and $a_2(t)$ in the expression for

$$e^{At} = a_1(t)I_2 + a_2(t)A$$

are calculated as

$$a_1(t) = 2e^{-t} - e^{-2t}, \quad a_2(t) = e^{-t} - e^{-2t}$$

Then

$$g_{11} = 11/12, \quad g_{12} = g_{21} = 1/4, \quad g_{22} = 1/12$$

Since $M = [\Gamma, A^T\Gamma] = I_2$, the solution matrix P is

$$P = MGM^T = G = \begin{bmatrix} \frac{11}{12} & \frac{1}{4} \\ \frac{1}{4} & \frac{1}{12} \end{bmatrix}$$

$$\triangle$$

The efficiency of this method depends on the efficiency in calculating the matrix G and the coefficients $a_i(t)$. If the eigenvalues of A are known a priori, for example in the case when A is a system matrix of a closed-loop pole placement problem, G can be calculated through relatively simple expression (Mori et al., 1986). Another advantage of this technique is that since G depends only on A, therefore for different $Q's$ (this case appears in practice for the Lyapunov equation corresponding to the power systems and large flexible structures, (Hodel, 1992)) equation (2.12) can be used over and over again without calculating G each time.

The presented approach due to (Mori et al., 1986) is also theoretically important, and it will be used in Section 2.3 in determining some of the bounds of the solution of the algebraic Lyapunov equation (2.1).

Note that an infinite series expansion for the solution of (2.1) is obtained in (Barnett and Storey, 1967c).

2.1.2 Skew-Symmetric Matrix Approach

For large systems where hundreds of variables are present, the order n of the matrix P can be very large. In such cases the number of equation for the solution of P may become very large. Thus, any effort to reduce the number of equations becomes useful. The actual number of simultaneous linear equations to be solved in order to get a solution of (2.1) can be reduced from $0.5n(n+1)$ to $0.5n(n-1)$ by introducing a skew-symmetric matrix S as follows, (Barnett and Storey, 1966a, 1966b; Barnett, 1974) — see also (Jacyno, 1989; Shen and Zelentsovsky, 1991)

$$PA - A^T P = S \tag{2.14}$$

Adding (2.1) and (2.14), we obtain

$$P = \frac{1}{2}(S - Q)A^{-1} \tag{2.15}$$

Using (2.15) and the facts that $P = P^T$, $S^T = -S$, the matrix S can be obtained as the solution of the following equation

$$A^T S + SA = A^T Q - QA \tag{2.16}$$

Since both S and the right-hand side of (2.16) are skew-symmetric matrices, (2.16) represents only $0.5n(n-1)$ linear equations and unknowns.

Example 2.4: Consider the third-order linear dynamical system where the system matrix is

$$A = \begin{bmatrix} 0 & 1 & -2 \\ 3 & -4 & 5 \\ -6 & 7 & 8 \end{bmatrix}$$

If we choose $Q = I_3$, then applying the previous method, we obtain the following sixth-order system of linear equations for the six unknown entries of the matrix P

$$\begin{bmatrix} 0 & 6 & -12 & 0 & 0 & 0 \\ 1 & -4 & 7 & 3 & -6 & 0 \\ -2 & 5 & 8 & 0 & 3 & -6 \\ 0 & 2 & 0 & -8 & 14 & 0 \\ 0 & -2 & 1 & 5 & 4 & 7 \\ 0 & 0 & -4 & 0 & 10 & 16 \end{bmatrix} \begin{bmatrix} p_{11} \\ p_{12} \\ p_{13} \\ p_{22} \\ p_{23} \\ p_{33} \end{bmatrix} = \begin{bmatrix} -1 \\ 0 \\ 0 \\ -1 \\ 0 \\ -1 \end{bmatrix}$$

which yields the following matrix P

$$P = \begin{bmatrix} p_{11} & p_{12} & p_{13} \\ p_{12} & p_{22} & p_{23} \\ p_{31} & p_{32} & p_{33} \end{bmatrix} = \begin{bmatrix} -0.9015 & -0.3077 & -0.0705 \\ -0.3077 & 0.0012 & -0.0268 \\ -0.0705 & -0.0268 & -0.0634 \end{bmatrix}$$

Using the skew-symmetric matrix S as in (2.16), we have the following third-order system of linear equations for the three unknowns of the matrix S

$$\begin{bmatrix} 5 & 8 & 3 \\ 4 & -7 & -6 \\ 2 & 1 & 4 \end{bmatrix} \begin{bmatrix} s_{12} \\ s_{13} \\ s_{23} \end{bmatrix} = \begin{bmatrix} -4 \\ -2 \\ 2 \end{bmatrix}$$

producing the following matrix S

$$S = \begin{bmatrix} 0 & s_{12} & s_{13} \\ -s_{12} & 0 & s_{23} \\ -s_{13} & -s_{23} & 0 \end{bmatrix} = \begin{bmatrix} 0 & -0.3286 & -0.6 \\ 0.3286 & 0 & 0.8143 \\ 0.6 & -0.8143 & 0 \end{bmatrix}$$

It can be easily verified that

$$P = 0.5(S - Q)A^{-1}$$

$$\triangle$$

2.1.3 Special Cases

Special case methods comprise a class of methods which cover the cases when the system matrix possesses one of the canonical forms: phase variable (companion), Schwarz, or Jordan form. It is important to point out that even though the calculations are drastically simplified, the use of canonical forms, in general, is coupled with the numerical ill-conditioning of the problems under consideration (Householder, 1964; Wilkinson, 1965). The original Lyapunov equation (2.1) can be converted into an equivalent form

$$B^T Y + Y B + R = 0 \tag{2.17}$$

where B, being in canonical form, is similar to A through the transformation matrix T, that is

$$B = T A T^{-1} \tag{2.18}$$

According to this transformation the relations between the original and new matrices are given by

$$Y = T^{-T} P T^{-1}, \quad R = T^{-T} Q T^{-1} \tag{2.19}$$

The new system (2.17) also has $0.5n(n+1)$ linear equations, but the transformation matrix T is chosen such that the coefficient matrix B for this new system (2.17) is in one of canonical forms and thus very sparse matrix, allowing us to perform less operations. Special cases are considered for the matrix B having companion (phase variable) canonical form, Schwarz form, and Jordan form.

Schwarz Form

The Schwarz form of a square matrix is defined by (Schwarz, 1956)

$$B = \begin{bmatrix} 0 & 1 & 0 & 0 & \ldots & 0 \\ -b_n & 0 & 1 & 0 & \ldots & 0 \\ 0 & -b_{n-1} & 0 & 1 & \ldots & 0 \\ \vdots & \vdots & \vdots & \vdots & \ddots & \vdots \\ 0 & 0 & 0 & \ldots & 0 & 1 \\ 0 & 0 & 0 & \ldots & -b_2 & -b_1 \end{bmatrix} \tag{2.20}$$

The elements of the Schwarz form (that is, $b_1, b_2, \ldots, b_n$) can be easily evaluated from the first column of the Routh array (Gantmacher, 1959) of the corresponding phase variable canonical form of the given matrix (Chen and Chu, 1966).

In the problem of finding the Lyapunov functions for examining the stability of linear systems, (Kalman and Bertram, 1960), it is shown that by choosing the matrix R as $R = diag[0, 0, \ldots, 0, 2b_1^2]$ the solution of (2.17) is easily obtained as

$$Y = diag[b_1 b_2 \cdots b_n, \; b_1 b_2 \cdots b_{n-1}, \; \ldots, \; b_1 b_2, \; b_1] \tag{2.21}$$

The algebraic Lyapunov equation for the system matrix in the Schwarz form and with the matrix R having all elements equal to zero except for $r_{nn} \neq 0$ was considered also in (Butchart, 1965). Similarly, the solution has been obtained for R having one nonzero entry in a general position on the principal diagonal (Barnett and Storey, 1968). In a series of papers by (Power 1967a, 1967b, 1967c, 1969) different structures for the matrix R of the algebraic Lyapunov equation in the Schwarz form and the corresponding methods for solving this equation have been considered.

Ziedan takes the advantage of the result given in (2.21) and establishes a relation between the solution of (2.17) (Y matrix) with the matrix B in the Schwarz form, and the solution of (2.1) (P matrix) for an arbitrary matrix A and with the matrix Q of the rank one. If we replace B in (2.17) by TAT^{-1}, we obtain (Ziedan, 1972)

$$A^T(T^TYT) + (T^TYT)A + T^TRT = 0 \qquad (2.22)$$

Since $R = diag[0, 0, \ldots, 0, 2b_1^2]$ and $T^TRT = 2b_1^2 t_n t_n^T$, where t_n^T is the n-th row of T, equation (2.22) becomes

$$\frac{1}{2b_1^2} A^T(T^TYT) + \frac{1}{2b_1^2}(T^TYT)A + t_n t_n^T = 0 \qquad (2.23)$$

It is shown in (Ziedan, 1972) that while choosing t_n arbitrary, we can easily obtain $t_1^T, t_2^T, \ldots, t_{n-1}^T$ (rows of T) recursively from equation

$$TA = BT$$

as

$$t_{n-1}^T = -\frac{1}{b_2} t_n^T(b_1 I + A)$$

$$\cdots \qquad (2.24)$$

$$t_{n-i}^T = \frac{1}{b_{i+1}}\left(t_{n-i+2}^T - t_{n-i+1}^T A\right), \quad i = 2, 3, \ldots, n-1$$

Assuming that Q is positive definite, it can be expressed as

$$Q = \Gamma\Gamma^T = \sum_{i=1}^{n} \gamma_i \gamma_i^T$$

where γ_i is the i-th column of the matrix Γ. It follows from (Ziedan, 1972) that the matrix P can be expressed in terms of T and Y as follows

$$P = \frac{1}{2b_1^2} \sum_{i=1}^{n-1} T_i^T Y T_i \qquad (2.25)$$

where T_i is the transformation matrix T with its n-th row chosen as γ_i^T, and then computing other rows recursively from equation (2.24). Equivalently, (2.25) can be written as

$$P = \frac{1}{2b_1^2} \sum_{i=0}^{n-1} J_i Q J_i y_{(n-i)(n-i)} \tag{2.26}$$

where

$$J_0 = I, \ J_1 = \frac{1}{b_2}(A + b_1 I), \ J_i = \frac{1}{b_{i+1}}(J_{i-2} + J_{i-1}A)$$
$$i = 2, 3, ..., n-1$$

and y_{ii} is the (i, i)-th element of the diagonal matrix Y.

This method expresses the solution as a sum of matrices and does not require calculation of any matrix inverse. Therefore, for higher order systems this method may be more efficient than the earlier described direct expansion methods.

Companion — Phase Variable Form

Algebraic Lyapunov equation having the system matrix in companion form (also known as phase variable canonical form) is considered in (Barnett and Storey, 1967a; Guidorzi, 1972, Sreeram and Agathoklis, 1991; Xiao et al., 1992; Curran 1993). In the following we use the notation of the original Lyapunov equation (2.1) instead of that in (2.17). The special forms of the matrices will be clear from the context.

The companion form of a matrix A is given by

$$A = \begin{bmatrix} 0 & 1 & 0 & 0 & \cdots & 0 \\ 0 & 0 & 1 & 0 & \cdots & 0 \\ \vdots & \vdots & \vdots & \vdots & \ddots & \vdots \\ -a_n & -a_{n-1} & -a_{n-2} & -a_{n-3} & \cdots & -a_1 \end{bmatrix} \tag{2.27}$$

where $a_i, i = 1, 2, ..., n$, are the coefficients of the characteristic polynomial of matrix A.

Assuming that the matrix Q has the form

$$Q = \{2q_{ij}\}$$

where

$$q_{ij} = 0, \quad n+i \quad or \quad n+j \quad odd$$

$$q_{ij} = a_{n+1-i}a_{n+j-1}, \quad n+i \quad and \quad n+j \quad even$$

it has been shown in (Barnett and Storey, 1967a) that the solution of (2.1) with A given by (2.27) can be obtained directly as follows

$$P = \{p_{ij}\}$$

where

$$p_{ij} = 0, \quad i+j \quad odd$$

$$p_{ij} = \sum_{k=r}^{i-1}(-1)^{k+i-1}a_{n-k}a_{n-i-j+1+k}, \quad i+j \quad even$$

with

$$r = 0, \quad if \quad i+j \le n+1$$

$$r = i+j-n-1, \quad if \quad i+j > n+1$$

Example 2.5: Consider the following matrix in phase variable canonical form

$$A = \begin{bmatrix} 0 & 1 & 0 \\ 0 & 0 & 1 \\ -2 & -9 & -8 \end{bmatrix}$$

In order to obtain the matrix B in Schwarz form, which is similar to A, we construct the Routh table

s^3	1	9
s^2	8	2
s	8.75	0
s^0	2	0

Using the elements of the Routh column (first column) the matrix B in the Schwarz form is obtained as (Chen and Chu, 1966)

$$B = \begin{bmatrix} 0 & 1 & 0 \\ -0.25 & 0 & 1 \\ 0 & -8.75 & -8 \end{bmatrix}$$

Thus, we have $b_1 = 8, b_2 = 8.75, b_3 = 0.25$. By choosing

$$R = diag\{0,\ 0,\ 128\}$$

we get the solution of the algebraic Lyapunov equation in Schwarz form (2.17) from (2.21) as

$$Y = \begin{bmatrix} 17.5 & 0 & 0 \\ 0 & 70 & 0 \\ 0 & 0 & 8 \end{bmatrix}$$

Δ

Example 2.6: Consider the same matrix A in phase variable canonical form as in Example 2.5, that is, with $a_3 = 2, a_2 = 9, a_1 = 8, a_0 = 1$. If the matrix Q in the algebraic Lyapunov equation (2.1) is chosen according to (Barnett and Storey, 1967a) as explained above, that is

$$Q = \begin{bmatrix} 8 & 0 & 32 \\ 0 & 0 & 0 \\ 32 & 0 & 128 \end{bmatrix}$$

then the solution of this Lyapunov equation is given by

$$p_{11} = a_3 a_2 = 18, \quad p_{12} = p_{21} = p_{23} = p_{32} = 0$$
$$p_{13} = a_3 a_0 = 2, \quad p_{22} = -a_3 a_0 + a_2 a_1 = 70$$
$$p_{31} = a_3 a_0 = 2, \quad p_{33} = a_1 a_0 = 8$$

Thus

$$P = \begin{bmatrix} 18 & 0 & 2 \\ 0 & 70 & 0 \\ 2 & 0 & 8 \end{bmatrix}$$

Δ

Guidorzi has studied a more general problem with the matrix Q being symmetric and arbitrary. He has presented a systematic procedure how to convert the algebraic Lyapunov equation (2.1) with A being in companion form (2.27) into a system of linear algebraic equations with the system matrix being sparse such that it can be easily triangularized (Guidorzi, 1972).

The problem of solving the algebraic Lyapunov equation in companion form has become recently an attractive research area (Sreeram and Agathoklis, 1991; Xiao et al., 1992; Curran, 1993).

Very interesting result is obtained by (Sreeram and Agathoklis 1991), namely, the solution of (2.1) subject to (2.27) for the case of an asymptotically stable matrix A is completely determined in terms of the elements of the Routh table. This solution is derived for the so-called single-input single-output case for which the matrix Q is given by

$$Q = bb^T, \qquad b^T = \begin{bmatrix} 0 & 0 & \dots & 0 & 1 \end{bmatrix} \qquad (2.28)$$

The solution is presented in the form of the following algorithm.

Algorithm 2.3:

1) Compute the reciprocal characteristic polynomial of A defined by

$$\Delta_r(\lambda) = \lambda^n \Delta(\lambda), \qquad \Delta(\lambda) = det[\lambda I - A]$$

2) Form the Routh table for the reciprocal characteristic polynomial and identify the corresponding elements T_{ij} and denote the number of elements in the $(n - k)$-th row by m.

3) Compute diagonal elements of the matrix P from

$$p_{nn} = \frac{1}{2T_{n1}}$$

$$p_{n-k,n-k} = \frac{1}{T_{n-k,1}} \left\{ -\sum_{i=1}^{m-1} (-1)^i T_{n-k,i+1} p_{n-k+i,n-k+i} \right\}$$

$$k = 1, 2, ..., n-1$$

4) Compute the off-diagonal elements of the matrix P as

$$p_{jk} = 0, \quad for \quad j + k \quad odd$$
$$q_{jk} = (-1)^{(k-j)/2} q_{ii}, \quad i = (j+k)/2, \quad j + k \quad even$$

$$\triangle$$

Another interesting approach for solving (2.1) subject to (2.27)-(2.28), (Xiao, 1992), computes also first diagonal elements of the matrix P, this time from a system of n linear algebraic equations whose coefficient matrix is given in terms of the coefficients of the characteristic polynomial of A, and then determines off-diagonal elements of P by using simple algebraic relations.

In the paper by (Curran, 1993) the frequency domain technique is developed for solving (2.1), (2.27)-(2.28). The method does not pretend to be numerically efficient, but it produces a solution even in the case when the basic existence and uniqueness condition $(\lambda_i + \lambda_j \neq 0)$ is not satisfied. Thus, this method can be used to find a solution of the algebraic Lyapunov equation in companion form.

Jordan Form

The Jordan form of an $n \times n$ matrix A is represented by the following block diagonal matrix

$$A = diag\left\{ J_1^{k_1}(\lambda_1), J_2^{k_2}(\lambda_2), ..., J_m^{k_m}(\lambda_m) \right\} \qquad (2.29)$$

where $\lambda_1, \lambda_2, ..., \lambda_m$ are distinct eigenvalues of matrix A of multiplicity $k_1, k_2, ..., k_m$ with $\sum_{i=1}^{m} k_i = n$ and $J_i^{k_i}(\lambda_i), \quad i = 1, 2, ..., m$ are $k_i \times k_i$ block diagonal matrices whose entries are Jordan blocks. Each Jordan block has on the main diagonal the corresponding eigenvalue and if the dimension of the block is greater than one above the main diagonal are ones. All other entries in the Jordan block are zeros. The complete theory of Jordan forms is complex and it can be found in several books on linear algebra, for example (Gantmacher, 1959; Lancaster and Tismenetsky, 1985).

The solution of (2.1) with the matrix A given by (2.29) can be obtained from the results of (Rutherford, 1932; Ma, 1966). As a matter of fact they studied the Sylvester equation (2.6) in the Jordan form — Rutherford by expanding it into a set of linear equations and Ma by using a finite series method. Since the Sylvester equation is more general than the Lyapunov equation the corresponding results for the Lyapunov equation can be obtained in a straightforward way by setting $A_1 = A_2 = A$. It is interesting to point out that the results of (Ma, 1966) are applicable even in the case when the matrix A is singular.

2.2 Solution Bounds

Finding the solution of the Lyapunov matrix equation is important in most of the applications, but for certain occasions we are just interested in the general behavior of the underlying system. This general behavior can be determined by examining only the partial or approximate solution, or just by examining certain bounds on the parameters of the solution instead of the full solution. Among these bounds the important ones are eigenvalue bounds, trace bounds, determinant bounds, geometric mean bounds, and arithmetic mean bounds. For example, the eigenvalue bounds can be used towards the design of suboptimal controller or to establish an effective bound on the settling time of the time-invariant system $\dot{x}(t) = Ax(t)$, $x(t_0) = x_0$. Trace and determinant bounds give rough estimates of the 'mean size' of the solution, which is helpful in numerical solutions. Also, the eigenvalues bounds can be used to construct the approximate solutions needed for the iterative methods for solving large scale algebraic Lyapunov equations. Eigenvalue bounds can be also used to determine whether or not the system under consideration possesses the singularly perturbed structure. Singularly perturbed differential and algebraic Lyapunov equations will be considered in Chapters 4 and 5.

This section describes important bounds on different parameters of the solution matrix P. We present the main results, analysis, and comparison where possible. Also some short proofs with the lemmas and basic inequalities used to establish these proofs are given in this chapter. For the complete proofs of other important results the reader is referred to the source papers. It is important to emphasize that despite many results on the bounds of the main attributes of the solution matrix P of the algebraic Lyapunov equation the best possible bounds have not been obtained yet so that the results available in the literature are pretty conservative.

2.2.1 Eigenvalue Bounds

Eigenvalue bounds play a key role in determining various properties of the solution matrix P. Different theorems that relate the eigenvalues of the matrices A, P, and Q are described in this section with brief analysis.

A result regarding the existence of the solution of (2.1) in the region defined by the condition that $P < Q$, $(P - Q$ is negative definite), in terms of the eigenvalues of the matrix A is given in (Man, 1969), which is corrected in (Barnett and Man, 1970; Peng, 1972). The necessary conditions are given in the following theorem.

Theorem 2.1 *The necessary conditions for the existence of a symmetric positive definite matrix P satisfying $A^T P + PA + 2\sigma Q = 0$, where Q is any symmetric positive definite matrix and σ is some positive scalar, for which $P < Q$, are:*
a) the symmetric matrix $(A + A^T + 2\sigma I)$ is negative definite, and
b) the real parts of all the eigenvalues of matrix A are less than $-\sigma$.

$\square$

It can be seen that the result given by Theorem 2.1 is limited to a subset of stable matrices A. The following result holds for all stable matrices A and all symmetric positive definite matrices Q, (Barnett and Man, 1970; Peng, 1972).

Theorem 2.2 *The sufficient condition for the existence of a symmetric positive definite matrix P satisfying $A^T P + PA + 2\sigma Q = 0$, where Q is any symmetric positive definite matrix and σ is some positive scalar, for which*

$$\min_{i} [\lambda_i(P - Q)] < 0, \quad i = 1, 2, \ldots, n$$

is that the real parts of all the eigenvalues of the matrix A are less than $-\sigma$.

$\square$

A relation between the largest and smallest eigenvalues of P and Q in (2.1) is derived in several papers (Shapiro, 1974; Montemayor and Womack, 1975; Kwon and Pearson, 1977; Fahmy and Hanafy, 1981). These bounds have been applied towards the design of suboptimal controllers for the minimum time problem and also may be used to establish an effective bound on the settling time of a time-invariant system and obtain information about the margins for system stability (Kalman and Bertram, 1960; Shapiro, 1972).

Theorem 2.3   *Let $\alpha_1, \alpha_2, \cdots, \alpha_n$ be the eigenvalues of P, $\beta_1, \beta_2, \cdots, \beta_n$ be the eigenvalues of Q, $\lambda_1, \lambda_2, \cdots, \lambda_n$ be the eigenvalues of A and $\sigma_1, \sigma_2, \cdots, \sigma_n$ be the eigenvalues of $A^T A$. It is assumed that P and Q are positive definite matrices. Using the ordering as*

$$0 < \alpha_n \leq \alpha_{n-1} \leq \cdots \leq \alpha_1$$

$$0 < \beta_n \leq \beta_{n-1} \leq \cdots \leq \beta_1$$

$$0 < \sigma_n \leq \sigma_{n-1} \leq \cdots \leq \sigma_1$$

and

$$Re[\lambda_n(A)] \leq Re[\lambda_{n-1}(A)] \leq \cdots \leq Re[\lambda_1(A)] < 0$$

the following inequalities hold

$$\lambda_{max}(P) = \alpha_1 \geq \frac{\beta_1}{2\sigma_1^{1/2}} = \frac{\lambda_{max}(Q)}{2\lambda_{max}^{1/2}(A^T A)}$$

$$\lambda_{min}(P) = \alpha_n \geq \frac{\beta_n}{2\sigma_1^{1/2}} = \frac{\lambda_{min}(Q)}{2\lambda_{min}^{1/2}(A^T A)}$$

$\square$

Example 2.7: Consider the fifth-order industrial reactor (Arkun and Ramakrishnan, 1983) where

$$A = \begin{bmatrix} -16.11 & -0.39 & 27.2 & 0 & 0 \\ 0.01 & -16.99 & 0 & 0 & 12.47 \\ 15.11 & 0 & -53.6 & -16.57 & 71.78 \\ -53.36 & 0 & 0 & -107.2 & 232.11 \\ 2.27 & 60.1 & 0 & 2.273 & -102.99 \end{bmatrix}$$

If we choose $Q = I_5$, then the solution of the Lyapunov matrix equation (2.1) is given by

$$P = \begin{bmatrix} 0.1423 & 0.0883 & 0.0710 & -0.0122 & 0.0303 \\ 0.0883 & 0.2595 & 0.0512 & -0.0043 & 0.0656 \\ 0.0710 & 0.0512 & 0.0453 & -0.0064 & 0.0206 \\ -0.0122 & -0.0043 & -0.0064 & 0.0057 & 0.0026 \\ 0.0303 & 0.0656 & 0.0206 & 0.0026 & 0.0330 \end{bmatrix}$$

The eigenvalues of A are

$$\lambda_1 = -2.94, \lambda_2 = -8.82, \lambda_3 = -74.44, \lambda_4 = -82.20, \lambda_5 = -128.50$$

The eigenvalues of $A^T A$ are

$$\sigma_5 = 3.97, \ \sigma_4 = 54.47, \ \sigma_3 = 4185.27, \ \sigma_2 = 5547.94, \ \sigma_1 = 82623.69$$

The bounds in Theorem 2.3, then yield

$$\lambda_{min}(P) = \alpha_5 \geq \frac{1}{2\sigma_1^{1/2}} = 0.00174$$

$$\lambda_{max}(P) = \alpha_1 \geq \frac{1}{2\sigma_1^{1/2}} = 0.00174$$

The exact values for the eigenvalues of matrix P are

$$\alpha_5 = 0.0026, \ \alpha_4 = 0.0072, \ \alpha_3 = 0.0172, \ \alpha_2 = 0.1115, \ \alpha_1 = 0.3473$$

Note that the obtained bound for α_5 is quite sharp, but the bound for α_1 is not very tight.

$$\Delta$$

Both upper and lower bounds for the minimal and maximal eigenvalues of the solution of the algebraic Lyapunov equation are obtained in (Yasuda and Hirai, 1979) as a by product of the more general study involving the corresponding bounds for the solution of the algebraic Riccati equation

$$A^T P + PA + Q - PRP = 0 \tag{2.30}$$

which degenerates to the algebraic Lyapunov equation for $R = 0$. The required bounds are obtained as

$$\frac{-1}{\min\left\{Re\{\lambda\{(A + A^T)Q^{-1}\}\}\right\}} \leq \lambda_{min}(P) \leq \frac{-\lambda_{max}(Q)}{2\min\left\{Re\{\lambda(A)\}\right\}}$$

(2.31a)

$$\frac{-\lambda_{min}(Q)}{2\max\left\{Re\{\lambda(A)\}\right\}} \leq \lambda_{max}(P) \leq \frac{-1}{\max\left\{Re\{\lambda\{(A + A^T)Q^{-1}\}\}\right\}}$$

$$valid \quad when \quad Re\{\lambda\{(A + A^T)Q^{-1}\}\} < 0$$

(2.31b)

Since these results are obtained in the context of a more general study the proof is omitted.

Note that bounds (2.31a) and (2.31b) are very similar to the bounds obtained by (Smith, 1965), which are summarized in the following theorem.

Theorem 2.4 Let $A < 0$ and $Q > 0$. Then, the minimal and maximal eigenvalues of the solution matrix P of (2.1) satisfy

$$\frac{\lambda_{min}(Q)}{|\lambda_{min}(A + A^T)|} \leq \lambda_{min}(P) \leq \frac{\lambda_{max}(Q)}{2|\min\left\{Re\{\lambda(A)\}\right\}|}$$

(2.31c)

$$\frac{\lambda_{min}(Q)}{2|\max\left\{Re\{\lambda(A)\}\right\}|} \leq \lambda_{max}(P) \leq \frac{\lambda_{max}(Q)}{|\lambda_{max}(A + A^T)|}$$

(2.31d)

$$valid \quad when \quad \lambda_{max}(A + A^T) < 0$$

$\square$

Proof: An elegant proof for these inequalities is given in (Lancaster, 1970). It is obtained as follows. Multiply the algebraic Lyapunov equation (2.1) from the left and right by the eigenvectors of A, that is

$$x^*\left(A^T P + PA\right)x = -x^*Qx$$

where $*$ indicates the complex conjugate transpose. Then, we have

$$2Re\{\lambda(A)\}x^*Px = -x^*Qx$$

By the extremal properties of the Rayleigh quotient, it follows

$$\lambda_{min}(P) \leq \frac{\lambda_{max}(Q)}{2|\min\left\{Re\{\lambda(A)\}\right\}|}, \quad \lambda_{max}(P) \geq \frac{\lambda_{min}(Q)}{2|\max\left\{Re\{\lambda(A)\}\right\}|}$$

The second parts of the inequalities (2.31c) and (2.31d) are established by using the argument that for $\mu > 0$ and a vector $y \neq 0$ the relation $Py = \mu y$ transforms (2.1) into

$$\mu y^* \left(A + A^T \right) y = -y^* Q y$$

Choosing μ as $\mu = \lambda_{max}(P)$ the last equation implies the upper bound given in (2.31d). The choice of μ as $\mu = \lambda_{min}(P)$ implies the lower bound in (2.31c). Note that the upper bound in (2.31d) is valid when $\lambda_{max}\left(A + A^T\right) < 0$, (Smith, 1965). ∎

Example 2.8: Consider a mathematical model of a voltage regulator whose system matrix is given by (Kokotovic, 1972)

$$A = \begin{bmatrix} -0.2 & 0.5 & 0 & 0 & 0 \\ 0 & -0.5 & 1.6 & 0 & 0 \\ 0 & 0 & -14.28 & 85.71 & 0 \\ 0 & 0 & 0 & -25 & 75 \\ 0 & 0 & 0 & 0 & -10 \end{bmatrix}$$

Using MATLAB we have obtained

$$\lambda_{min}\left(A + A^T\right) = -152.0743, \quad \lambda_{max}\left(A + A^T\right) = 77.2679$$
$$\min\left\{Re\{\lambda(A)\}\right\} = -25, \quad \max\left\{Re\{\lambda(A)\}\right\} = -0.2$$

$$\lambda_{min}(P) = 0.0110, \quad \lambda_{max}(P) = 34.8851$$

By Theorem 2.4 we have

$$0.0066 \leq \lambda_{min}(P) \leq 0.02, \quad 2.5 \leq \lambda_{max}(P)$$

Note that since $\lambda_{max}\left(A + A^T\right)$ is positive, the upper bound in (2.31d) is not applicable. From our computational experience with general system matrices, the condition $\lambda_{max}\left(A + A^T\right) < 0$ is satisfied only in rare cases of diagonally dominant stable matrices and symmetric stable matrices so that the upper bound from (2.31d) hardly can be obtained. The fact that $Q = I$ implies exactly the same bounds from (2.31a) and (2.31b). It is interesting to compare bounds in (2.31) when the matrix $Q \neq I$.

Consider the following matrix

$$Q_1 = \begin{bmatrix} 4 & 1 & 0 & 2 & 0 \\ 1 & 10 & 3 & -1 & 0 \\ 0 & 3 & 5 & 1 & 0 \\ 2 & -1 & 1 & 10 & 0 \\ 0 & 0 & 0 & 0 & 6 \end{bmatrix} > 0$$

For this value of Q_1 we get

$$\lambda_{max}(Q_1) = 11.5855, \quad \lambda_{min}(Q_1) = 2.5572$$
$$\min\left\{Re\left\{\lambda\left\{(A + A^T)Q_1^{-1}\right\}\right\}\right\} = -30.3328$$
$$\max\left\{Re\left\{\lambda\left\{(A + A^T)Q_1^{-1}\right\}\right\}\right\} = 10.2374$$

The actual extremal eigenvalues are

$$\lambda_{min}(P_1) = 0.0544, \quad \lambda_{max}(P_1) = 233.569$$

Both results from (2.31) imply the same bounds for the maximal eigen-value of P_1, but the results of (Yasuda and Hirai, 1979) appear to be sharper for the minimal eigenvalue in this particular example. Namely, we get from (2.31a)

$$0.03 \leq \lambda_{min}(P_1) \leq 0.075$$

and from (2.31c)

$$0.0168 \leq \lambda_{min}(P_1) \leq 0.075$$

$\triangle$

Theorems 2.3–2.4 and formula (2.31) present bounds for the extreme eigenvalues of the solution matrix P in terms of the eigenvalues and singular values of matrices A and Q. The bounds concerning all the eigenvalues of the matrix P are obtained by (Karanam, 1981).

Theorem 2.5 *With the notation from Theorem 2.3 the following inequalities hold*

$$\alpha_j \geq \frac{\beta_{j+k-1}}{2\sigma_k^{1/2}}, \quad j, k \geq 1 \quad and \quad j + k \leq n + 1$$

$\square$

Proof of this theorem is based on the following inequalities known from (Fan and Hoffman, 1955; Fan, 1951).

Fan-Hoffman's Inequality:

For any square $n \times n$ matrix Z the following holds

$$\lambda_i\left(\frac{Z + Z^T}{2}\right) \leq \sigma_i(Z), \quad 1 \leq i \leq n \tag{2.32}$$

Fan's Singular Values Inequalities:

For any two square $n \times n$ matrices X, Y the singular values satisfy

$$\begin{aligned} i) \quad & \sigma_{j+k-1}(XY) \leq \sigma_j(X)\sigma_k(Y), \quad j, k \geq 1, \ j + k \leq n + 1 \\ ii) \quad & \sigma_{j+k-1}(X + Y) \leq \sigma_j(X) + \sigma_k(Y), \quad j, k \geq 1, \ j + k \leq n + 1 \end{aligned} \tag{2.33}$$

Proof: The proof of Theorem 2.5 now can be completed as follows. Setting $Z = -2PA$ in (2.32) we get from (2.1)

$$\lambda_i(-PA - A^T P) = \lambda_i(Q) \leq \sigma_i(-2PA)$$

Choosing $X = 2PA$, $Y = -A$ in (2.33) implies

$$\lambda_{j+k-1}(Q) \leq \sigma_j(2P)\sigma_k(-A) = 2\alpha_j \sigma_k^{1/2}$$

which according to the established notation proves Theorem 2.5 since $\beta_{j+k-1} = \lambda_{j+k-1}(Q)$.

■

Note that the obtained result in Theorem 2.5 is very general. For $k = 1$, we have

$$\alpha_j \geq \frac{\beta_j}{2\sigma_1}, \quad j = 1, 2, ..., n \tag{2.34}$$

and for $j = 1, n$ we get inequalities from Theorem 2.3. In certain cases a proper choice of k can yield tighter bounds, for example, take $Q = I$, $k = n, j = 1$, which yields to $\alpha_1 \geq 0.5/\sigma_n^{1/2}$, and since $\sigma_n \leq \sigma_1$ this gives tighter bound as compared to the first bound in Theorem 2.3. In the following example by choosing $k = n - j + 1$ the sharper bounds are obtained than in Example 2.7.

Example 2.9: As an illustration, for $k = n - j + 1$, the bounds for the eigenvalues of matrix P from Example 2.7 are calculated as

$$\alpha_1 \geq 0.2509, \ \alpha_2 \geq 0.0677, \ \alpha_3 \geq 0.0077, \ \alpha_4 \geq 0.0067, \ \alpha_5 \geq 0.0017$$

while the exact eigenvalues are

$$\alpha_1 = 0.3473, \ \alpha_2 = 0.1115, \ \alpha_3 = 0.0172, \alpha_4 = 0.0072, \ \alpha_5 = 0.0026$$

It can be seen that the bound for α_1 is now much sharper than in Example 2.7.

$\triangle$

Results of (Karanam, 1981) can also be used to estimate bounds for the trace and determinant of the solution of the algebraic Lyapunov equation (2.1), which will be discussed in the following subsections.

Next theorem first presents a majorization result relating eigenvalues of A and $P^{-1}Q$, and then gives bounds for extremal eigenvalues of P as a by product.

Theorem 2.6 (*Karanam, 1982; Wimmer, 1975*). *For $m = 1, 2, ..., n$, we have*

$$-\sum_{i=1}^{m} 2Re[\lambda_{n-i+1}(A)] \leq -\sum_{i=1}^{m} \lambda_{n-i+1}(-P^{-1}Q)$$

where equality holds for $m = n$. Furthermore, let

$$q_m = \sum_{j=1}^{m} \lambda_j(Q), \qquad q_{-m} = \sum_{j=1}^{m} \lambda_{n-j+1}(Q)$$

$$a_m = -\sum_{j=1}^{m} Re[\lambda_j(A)] \quad and \quad a_{-m} = -\sum_{j=1}^{m} Re[\lambda_{n-j+1}(A)]$$

then the following inequalities are satisfied

$$(i) \quad \lambda_1(P) \geq \frac{q_{-m}}{2a_m}, \qquad m = 1, 2, \dots, n$$

$$(ii) \quad \lambda_n(P) \leq \frac{q_m}{2a_{-m}}, \qquad m = 1, 2, \dots, n$$

$\square$

In (Karanam, 1982, 1986) the following bounds for all of the eigenvalues are also obtained

$$\frac{1}{\lambda_{n-j}(P)} \geq \frac{2a_{-m} - (q_j/\lambda_n(P))}{q_k}$$

$$\lambda_{j+1}(P) \geq \frac{q_{-k}}{2a_m - (q_{-j}/\lambda_1(P))}$$

$$j = 1, 2, ..., n - 1; \quad k = 1, 2, ..., n - j; \quad m = j + k$$

(2.35)

The lower bound for all eigenvalues is obtained in a simpler form in (Karanam, 1983) as

$$\lambda_i(P) \geq \frac{\beta_j}{2\sigma_{j-i+1}^{1/2}}, \quad n \geq j \geq i \geq 1 \tag{2.36}$$

with β_j and σ_j defined in Theorem 2.3.

The proof of the results of (Karanam, 1981, 1982, 1983) is omitted due to the fact that these results are obtained as special cases of the more general study done for the bounds of the solution's attributes of the algebraic Riccati equation.

Note that for $m = n$, inequality (i) was obtained by (Patel and Toda, 1978), which can be rewritten as

$$\lambda_1(P) \geq \frac{tr(Q)}{-2tr(A)} \tag{2.37}$$

with $\lambda_1(P) = \lambda_{max}(P)$ representing the spectral norm of P.

While the above theorems, in some cases, can give trivial bounds, the following theorem always generates nontrivial bounds (Mori et al., 1986). However, the solution is given in terms of the controllability matrix (2.11) so that it is computationally involved.

Theorem 2.7 *Following the notation of (2.11)-(2.13) and denoting by $\lambda_i(X)$ the i-th eigenvalue of X, the following inequalities hold*

$$i) \quad \lambda_{min}(G)MM^T \leq P \leq \lambda_{max}(G)MM^T$$

$$ii) \ \lambda_{min}(G)\lambda_i(MM^T) \leq \lambda_i(P) \leq \lambda_{max}(G)\lambda_i(MM^T), \ i = 1, 2, \ldots, n$$

where M is the controllability matrix of the pair (A, Γ).

□

Proof: From (2.13) and the properties of the Kronecker product we have

$$\lambda_1(H) = \lambda_1(G), \qquad \lambda_n(H) = \lambda_n(G)$$

Applying the following matrix inequalities

$$\lambda_n(X)YY^T \leq YXY^T \leq \lambda_1(X)YY^T, \qquad X = X^T$$

to (2.12) with $X = H$ and $M = G$, the first pair of inequalities of Theorem 2.7 is obtained. Then, part ii) follows directly from part i).

■

Example 2.10: Consider the same matrix A as in Example 2.3, but take different matrix Γ, that is

$$A = \begin{bmatrix} 0 & 1 \\ -2 & -3 \end{bmatrix}, \qquad \Gamma = \begin{bmatrix} 1 \\ 2 \end{bmatrix}$$

Then

$$M = [\Gamma, A^T\Gamma] = \begin{bmatrix} 1 & -4 \\ 2 & -5 \end{bmatrix}, \qquad MM^T = \begin{bmatrix} 17 & 22 \\ 22 & 29 \end{bmatrix}$$

The matrix G was calculated in Example 2.3 as

$$G = \begin{bmatrix} 11/12 & 1/4 \\ 1/4 & 1/12 \end{bmatrix}$$

with $\lambda_{min}(G) = 0.014$ and $\lambda_{max}(G) = 0.986$. The solution matrix P is given by

$$P = \begin{bmatrix} 1/4 & 1/4 \\ 1/4 & 3/4 \end{bmatrix}$$

with $\lambda_1(P) = 0.854$ and $\lambda_2(P) = 0.146$. According to Theorem 2.7, the bounds are

$$0.6413 \leq \lambda_1(P) \leq 45.1627$$

$$0.002744 \leq \lambda_2(P) \leq 0.1932$$

△

The nontrivial bounds given in Theorem 2.7 can be improved and computations considerably reduced in the case when A has multiple eigenvalues as given in (Troch, 1987). Here the reduced form of the controllability matrix M is used.

Theorem 2.8 *Let the number of distinct eigenvalues of A be $m \leq n$. Define $\hat{M}$ as*

$$\hat{M} = [\Gamma, A^T\Gamma, \ldots, (A^T)^{m-1}\Gamma]$$

where Γ is defined as $Q = \Gamma^T\Gamma$, and define $\hat{G}$ the same way as in (2.13) except that it is of dimension $m \times m$ instead of $n \times n$, then

$$i) \quad \lambda_{min}(\hat{G})\hat{M}\hat{M}^T \leq P \leq \lambda_{max}(\hat{G})\hat{M}\hat{M}^T$$

$$ii) \ \lambda_{min}(\hat{G})\lambda_i(\hat{M}\hat{M}^T) \leq \lambda_i(P) \leq \lambda_{max}(\hat{G})\lambda_i(\hat{M}\hat{M}^T), i = 1, 2, \ldots, n$$

$\square$

The usefulness of Theorem 2.8 is observed since we only have to factor minimal polynomial for $\lambda(\hat{G})$ as opposed to the characteristic polynomial for $\lambda(G)$, hence requiring less computations.

The generalization of the lower bounds of the eigenvalues of P in terms of the eigenvalues of A and Q is given in (Komaroff, 1988). In this very comprehensive paper both the lower summation bound and lower product bound are obtained. The main results are given in the following theorem.

Theorem 2.9 *For the solution of (2.1), the following inequalities hold*

$$i) \quad \prod_{i=1}^{k} \lambda_i(P) \geq \prod_{i=1}^{k} \lambda_{n-i+1}(Q) \prod_{i=1}^{k} [-2Re\{\lambda_i(A)\}]^{-1}, \ k = 1, 2, \ldots, n$$

$$ii) \sum_{i=1}^{k} \lambda_i(P) \geq k \left[\prod_{i=1}^{k} \lambda_{n-i+1}(Q) \right]^{\frac{1}{k}} \prod_{i=1}^{k} \frac{1}{[-2Re\{\lambda_i(A)\}]^{\frac{1}{k}}}$$
$$k = 1, 2, \ldots, n$$

$\square$

The proof of part i) of Theorem 2.9 is a direct consequence of the inequalities due to Horn and Fan.

Horn's Inequality:

For any two real symmetric positive semi-definite matrices we have

$$\prod_{i=1}^{k} \lambda_i(XY) \leq \prod_{i=1}^{k} \lambda_i(X)\lambda_i(Y), \quad k = 1, 2, ..., n \quad (2.38a)$$

with equality for $k = n$, (Horn, 1950). The second part of Horn's inequality is given by

$$\prod_{i=1}^{k} \lambda_{n-i+1}(XY) \geq \prod_{i=1}^{k} \lambda_{n-i+1}(X)\lambda_{n-i+1}(Y), \quad k = 1, 2, ..., n$$

$$(2.38b)$$

Fan's Eigenvalue Product Inequality:

For an $n \times n$ matrix X, representing a linear transformation in an unitary space, if $\lambda_i(X^T + X) \geq 0$ then

$$\prod_{i=1}^{k} \lambda_{n-i+1}(X^T + X) \leq \prod_{i=1}^{k} 2Re\{\lambda_{n-i+1}(X)\}, \quad k = 1, 2, ..., n$$

$$(2.39)$$

(Fan, 1953).

Proof: Using the above inequalities Theorem 2.9 can be proved as follows. Rewrite (2.1) as

$$-P^{-1/2}A^T P^{1/2} - P^{1/2}AP^{-1/2} = P^{-1/2}QP^{1/2}$$

Then

$$-\lambda_i\left(P^{-1/2}A^T P^{1/2} + P^{1/2}AP^{-1/2}\right)$$
$$= \lambda_i\left(P^{-1/2}QP^{1/2}\right) = \lambda_i\left(P^{-1}Q\right)$$

Using the second form of the Horn inequality, (2.38b), it follows that

$$\prod_{i=1}^{k} \lambda_{n-i+1}\left(P^{-1}\right)\lambda_{n-i+1}(Q)$$

$$\leq \prod_{i=1}^{k} \lambda_{n-i+1}\left(-P^{-1/2}A^T P^{1/2} - P^{1/2}AP^{-1/2}\right)$$

which by Fan's inequality, (2.39), implies

$$\prod_{i=1}^{k} \lambda_{n-i+1}(P^{-1}) \lambda_{n-i+1}(Q) \leq \prod_{i=1}^{k} [-2Re(\lambda_{n-i+1}(A))]$$

By using the fact that

$$\lambda_{n-i+1}(P^{-1}) = \lambda_i^{-1}(P)$$

we get the proof for part i). Part ii) is obtained from part i) by applying the very well-known arithmetic-mean geometric-mean inequality (Mitrinovic, 1970)

$$\left(\prod_{i=1}^{n} x_i\right)^{\frac{1}{n}} \leq \frac{1}{n} \sum_{i=1}^{n} x_i, \qquad x_i \geq 0 \qquad (2.40)$$

■

Note that for $k = 1$ part i) of Theorem 2.9 yields

$$\lambda_1(P) \geq -\lambda_n(Q)[2Re\{\lambda_1(A)\}]^{-1}$$

which is the result obtained by (Yasuda and Hirai, 1979) and given in (2.31b). The second application of the geometric-mean arithmetic-mean inequality (2.40), this time to ii) of Theorem 2.9, produces

$$\sum_{i=1}^{k} \lambda_i(P) \geq -k^2 \left[\prod_{i=1}^{k} \lambda_{n-i+1}(Q)\right]^{\frac{1}{k}} \left[\sum_{i=1}^{k} 2Re\{\lambda_i(A)\}\right]^{-1} \qquad (2.41)$$

which can be simplified as

$$\sum_{i=1}^{k} \lambda_i(P) \geq -k^2 \lambda_{min}(Q) \left[\sum_{i=1}^{k} 2Re\{\lambda_i(A)\}\right]^{-1} \qquad (2.42)$$

Results for the lower summation bounds stated in (2.41)-(2.42) are important in their own right. In addition, they will be used in the subsequent subsection to give estimates for the trace of the matrix P.

In (Komaroff, 1990) the fundamental results on *the upper bounds* for the eigenvalues of the solution of the algebraic Lyapunov equation (2.1) are obtained. The results are very general and include both summation and product bounds so that they can be used to estimate single eigenvalues, trace, and determinant of the matrix P. They are presented in the following theorem.

Theorem 2.10 *Let the matrix P satisfies the algebraic Lyapunov equation (2.1) then*

$$\lambda_k(P) \le \left[\sum_{i=1}^{k} \lambda_i(Q)\right] \left[\sum_{i=1}^{k}(-1)\lambda_i(A^T + A)\right]^{-1}$$

$$k = 1, 2, ..., n, \quad \sum_{i=1}^{k} \lambda_i(A^T + A) < 0$$

and

$$\prod_{i=1}^{k} \lambda_i(P) \le \left[\sum_{i=1}^{k} \frac{1}{k}\lambda_i(Q)\right] \left[\prod_{i=1}^{k}(-1)\lambda_i(A^T + A)\right]^{-1}$$

$$\lambda_1(A^T + A) < 0, \quad k = 1, 2, ..., n$$

$\square$

For the complete proof of this important theorem see the original paper by Komaroff. Here, we only discuss the results. From this theorem it also follows that

$$\sum_{i=1}^{k} \lambda_i(P) \le (-1)\left[\sum_{i=1}^{k} \lambda_i(Q)\right][\lambda_1(A^T + A)]^{-1}, \ \lambda_1(A^T + A) < 0$$

$$k = 1, 2, ..., n$$

(2.43)

which is relaxed in the follow-up paper (Komaroff, 1992) into

$$\sum_{i=1}^{k} \lambda_i(P) \le (-1)\sum_{i=1}^{k} \lambda_i(Q)[\lambda_i(A^T + A)]^{-1}, \ \lambda_1(A^T + A) < 0$$

$$k = 1, 2, ..., n$$

(2.44)

Note that the result obtained in (Yasuda and Hirai, 1979), (2.31), is stronger than the result of Theorem 2.10 for $k = 1$. However, the work of (Komaroff, 1990) is more general since it gives results for any k. Also from the first part of Theorem 2.10, we have

$$\lambda_n(P) = \lambda_{min}(P) \leq -\frac{tr(Q)}{2tr(A)} \tag{2.45}$$

which is the result of (Karanam, 1982) for $k = n$. The results of Theorem 2.10 for the upper product bounds are complemented in (Komaroff, 1992) by the following inequality

$$\prod_{i=1}^{k} \lambda_{n-i+1}(P) \leq \left[\frac{1}{n}tr(Q)\right]^k \left[\prod_{i=1}^{k}(-1)\lambda_i\big(A^T + A\big)\right]^{-1} \tag{2.46}$$
$$\lambda_1\big(A^T + A\big) < 0, \quad k = 1, 2, ..., n$$

Example 2.11: Consider the eigenvalue bounds for the following problem

$$A = \begin{bmatrix} -2 & 0 & 1 \\ 0 & -2 & -1 \\ -1 & 0 & -1 \end{bmatrix}, \quad Q = I_3$$

The eigenvalues of the solution matrix P of the algebraic Lyapunov equation (2.1) are given by

$$\lambda_{max}(P) = \lambda_1(P) = 0.5461, \quad \lambda_2(P) = 0.2642$$
$$\lambda_3(P) = \lambda_{min}(P) = 0.2281$$

Using the upper bounds for the eigenvalues as given in Theorem 2.10, we get

$$\lambda_1 \leq 0.6306, \quad \lambda_2 \leq 0.3581, \quad \lambda_3 \leq 0.3000$$

Comparing the exact values and the obtained bounds we can conclude that these bounds are pretty sharp for this particular example.

$$\triangle$$

Some additional summation and product bounds can be also found in (Garloff, 1986; Hmamed, 1990).

Example 2.12: Let us apply the results of Theorem 2.10 to a real model of a synchronous machine connected to an infinite bus. Its system matrix is given by (Kokotovic et al., 1980)

$$A = \begin{bmatrix} -0.58 & 0 & 0 & -0.27 & 0 & 0.2 & 0 \\ 0 & -1 & 0 & 0 & 0 & 1 & 0 \\ 0 & 0 & -5 & 2.1 & 0 & 0 & 0 \\ 0 & 0 & 0 & 0 & 337 & 0 & 0 \\ -0.14 & 0 & 0.14 & -0.2 & -0.28 & 0 & 0 \\ 0 & 0 & 0 & 0 & 0 & 0.08 & 2 \\ -173 & 66.7 & -116 & 40.9 & 0 & -66.7 & -16.7 \end{bmatrix}$$

We have solved (2.1) with $Q = I_7$ and obtained the following eigenvalues for the solution matrix P

$$\lambda_1(P) = 3000.8278, \quad \lambda_2(P) = 14.6177$$
$$\lambda_3(P) = 3.2171, \quad \lambda_4(P) = 1.4435$$
$$\lambda_5(P) = 0.8919, \quad \lambda_6(P) = 0.1279, \quad \lambda_7(P) = 0.0271$$

The eigenvalues of the matrix $A + A^T$ are given by

$$\lambda_1(A + A^T) = 338.9509, \quad \lambda_2(A + A^T) = 78.3061$$
$$\lambda_3(A + A^T) = 0.1117, \quad \lambda_4(A + A^T) = -1.2382$$
$$\lambda_5(A + A^T) = -9.8848, \quad \lambda_6(A + A^T) = -113.1114$$
$$\lambda_7(A + A^T) = -340.0944$$

It can be seen that $\sum_{i=1}^{k} \lambda_i(A + A^T) < 0$ only for $k = 7$ so that we are able to get only the upper bound for $\lambda_7(P)$, which is given by

$$\lambda_7(P) \le 0.1491$$

Information about other upper bounds for the eigenvalues of P in this example cannot be obtained from Theorem 2.10.

$$\Delta$$

It is noted that different bounds can be applied for different situations and none of these results can be declared as the best for all the applications. Therefore, we have to test each bound for each specific case.

2.2.2 Trace Bounds

Like the eigenvalue bounds, trace bounds are also helpful in estimating various parameters of underlying problem. For example, trace of the solution of (2.1) gives the mean square error of the Kalman filter and the performance value of the Kalman regulator, therefore, knowledge of the bounds for the trace helps in analyzing the performance of the filters and regulators under various circumstances. Furthermore, such bounds immediately give rough estimates of the 'mean size' of the solutions so that they could be extremely helpful in numerical computations.

From the results of (Patel and Toda, 1978; Mori, 1985; Kwon et al., 1985; Kwon and Youn, 1986) the following theorem on lower bounds for the trace of the matrix P can be stated.

Theorem 2.11 *The trace of the positive definite solution matrix P in (2.1) satisfies the following inequalities*

$$(i) \quad tr(P) \geq -\frac{tr(Q)}{2tr(A)}, \quad (Patel \ and \ Toda, \ 1978)$$

$$(ii) \quad tr(P) \geq -\frac{n^2|Q|^{1/n}}{2tr(A)}, \quad (Mori, \ 1985)$$

$$(iii) \quad tr(P) \geq -\frac{\lambda_{min}(Q)n^2}{2tr(A)}, \quad (Kwon \ et \ al., \ 1985)$$

where $|Q|$ denotes the determinant of Q.

$\square$

It is difficult to compare these inequalities in general, but we can see that (i) is stronger than (ii) when $Q = kI$ and $k > 1$, but weaker when $k < 1$. Since A is assumed to be stable, therefore, the negative sign is cancelled by $tr(A)$.

Example 2.13: Consider the linearized mathematical model of an F-8 aircraft (Litkouhi, 1983) having

$$A = \begin{bmatrix} -0.015 & -0.0805 & -0.0011666 & 0 \\ 0 & 0 & 0 & 0.03333 \\ -2.28 & 0 & -0.84 & 1 \\ 0.6 & 0 & -4.8 & -0.49 \end{bmatrix}$$

Choosing $Q = I_4$, the solution matrix P is obtained as

$$P = \begin{bmatrix} 248.737 & 6.211 & -1.655 & -0.904 \\ 6.211 & 272.719 & -8.428 & 1.501 \\ -1.655 & -8.428 & 2.318 & -0.301 \\ -0.904 & 1.501 & -0.301 & 0.508 \end{bmatrix}$$

while $tr(P) = 524.282$. The bounds in Theorem 2.11 yield, respectively to

$$(i) \; tr(P) \geq 1.487$$

$$(ii) \; tr(P) \geq 5.948$$

$$(iii) \; tr(P) \geq 5.948$$

By choosing $Q_1 = diag\{1, 2, 3, 4\}$ we get $tr\{P_1\} = 1722.6$. In this case Theorem 2.11 produces

$$(i) \; tr\{P_1\} \geq 3.718, \;\; (ii) \; tr\{P_1\} \geq 13.166$$
$$(iii) \; tr\{P_1\} \geq 5.948$$

It can be noticed from this particular example that the trace bounds given in Theorem 2.11 are very conservative.

$$\Delta$$

The following bounds are obtained as by product of the bounds determined for the algebraic Riccati equation.

Theorem 2.12 *(Wang et al., 1986). Let $A_s = (A + A^T)/2$, then the positive definite matrix P satisfying (2.1) has the following lower and upper bounds*

$$i) \quad tr(P) \leq -tr(Q)/2\lambda_{max}(A_s) \quad if \;\; \lambda_{max}(A_s) < 0$$

$$ii) \quad tr(P) \geq -tr(Q)/2\lambda_{min}(A_s) \quad if \;\; \lambda_{min}(A_s) < 0$$

$$iii) \quad tr(P) \geq -tr(Q)/2tr(A).$$

$$\square$$

The following lemma, that will be used in the proof of Theorem 2.12, gives the trace bounds for a matrix product (Wang et al., 1986).

Lemma 2.1 *Let* $X, Y \in R^{n \times n}$ *be symmetric and* $Y \geq 0$, *then*

$$\lambda_{min}(X)tr(Y) \leq tr(XY) \leq \lambda_{max}(X)tr(Y)$$

$\square$

The relaxed forms of this important lemma are given in Appendix.

Proof: The proof of Theorem 2.12 uses Lemma 2.1 as follows. From (2.1) we have

$$tr\left(A^T P\right) + tr(PA) = -tr(Q)$$

Define $2A_s = A^T + A$, then

$$tr\left(A^T P\right) + tr(PA) = 2tr(PA_s) = -tr(Q)$$

By Lemma 2.1 we have

$$2tr(P)\lambda_{max}(A_s) \geq -tr(Q)$$

and

$$2tr(P)\lambda_{min}(A_s) \leq -tr(Q)$$

which completes the proof for parts (i) and (ii). The facts that $\lambda_{min}(A_s) > tr\{A_s\}$ for $\lambda_{min}(A_s) < 0$, and $2tr(A) = tr(A_s)$ comprise the proof for part (iii).

■

The following upper and lower trace bounds have been obtained in (Komaroff, 1988, 1992).

Theorem 2.13 *The matrices* P, A, *and* Q *in (2.1) satisfy*

$$-\sum_{i=1}^{n} \frac{\lambda_i(Q)}{\lambda_i(A^T + A)} \geq tr(P) \geq -\left[\sum_{i=1}^{n} \lambda_i^{1/2}(Q)\right]^2 \frac{1}{2tr(A)}$$
$$\lambda_1\left(A + A^T\right) < 0$$

$\square$

The proof of this theorem is lengthy so that the interested reader is referred to the original papers (Komaroff, 1988, 1992).

Example 2.14: Consider a real world example, a six-plate gas absorber (De Vlieger et al., 1982). The system matrix is given by

$$A = \begin{bmatrix} -1.173 & 0.6341 & 0 & 0 & 0 & 0 \\ 0.5390 & -1.173 & 0.6341 & 0 & 0 & 0 \\ 0 & 0.5390 & -1.173 & 0.6341 & 0 & 0 \\ 0 & 0 & 0.5390 & -1.173 & 0.6341 & 0 \\ 0 & 0 & 0 & 0.5390 & -1.173 & 0.6341 \\ 0 & 0 & 0 & 0 & 0.5390 & -1.173 \end{bmatrix}$$

The eigenvalues of the matrix A_s are all negative so that both lower and upper bounds from Theorem 2.12 can be applied. Using $Q = I$, it has been obtained that $1.3453 \leq tr(P) \leq 25.8509$. However, the real value for the trace of the solution matrix P is $tr\{P\} = 6.7809$. Thus, the obtained bounds are not very tight. Much better result is obtain for the same bounds by using Theorem 2.13, namely,

$$2.5575 \leq tr\{P\} \leq 6.8236$$

It can be seen that the upper bound is very sharp.

$\triangle$

As with the eigenvalue bounds nothing can be said absolute about the goodness of the different trace bounds. Different bounds can be used for different applications.

2.2.3 Determinant Bounds

Similarly to the trace bounds, the bounds for the determinant also give rough estimates of the mean size of the solution, and so are helpful for numerical computations. The first result on a bound of the determinant of the matrix P is given in (Bialas, 1980).

Theorem 2.14 *If $\lambda_i(A) + \lambda_j(A) \neq 0$, $i, j = 1, 2, ..., n$, and Q is a positive definite matrix, then the solution of (2.1) satisfies the inequality*

$$|P| \geq \frac{(-1)^n |Q|}{2^n |A|}$$

$\square$

Under the same assumptions as in Theorem 2.14 (Mori et al., 1981) have derived another lower bound for the determinant of P

$$|P| \geq |Q| \left(\frac{n}{-2tr(A)} \right)^n \qquad (2.47)$$

Also from the work of (Karanam, 1981) and his Theorem 2.5, it can be seen that for $k = 1$ the following lower bound can be obtained

$$|P| \geq \frac{|Q|}{2^n \sigma_1^{n/2}} \qquad (2.48)$$

with σ_1 defined in Theorem 2.3.

A stronger result is given in (Komaroff, 1988), which is obtained from Theorem 2.9, part (i), for $k = n$.

Theorem 2.15 *(Komaroff, 1988). The lower bound for the determinant of the matrix P is given by*

$$|P| \geq |Q| \prod_{i=1}^{n} [-2Re\{\lambda_i(A)\}]^{-1}$$

$\square$

The lower bounds from Theorem 2.15 is sharper than the ones given in Theorem 2.14 and (2.47).

Example 2.15: It is interesting to note that it is not possible to find a general upper bound, because $|P|$ may go to infinity as illustrated in the following example (Bialas, 1980). Let

$$A = \begin{bmatrix} sin\alpha & cos\alpha \\ -cos\alpha & sin\alpha \end{bmatrix}, \quad Q = \begin{bmatrix} 1 & 0 \\ 0 & 1 \end{bmatrix}.$$

Then for the solution of (2.1), we have

$$|P| = \frac{1}{4sin^2\alpha}$$

Therefore, $|P| \to \infty$ for $sin^2\alpha \to 0$ and $|A| = |Q| = 1$.

$\triangle$

An upper bound has been established recently in (Komaroff, 1990, 1992) as

$$|P| \leq \left[\frac{1}{n} tr(Q)\right]^n abs \left|A^T + A\right|^{-1}, \quad \lambda_1 \left(A^T + A\right) < 0 \qquad (2.49)$$

where $abs \left|A^T + A\right|$ stands for the absolute value of $\left|A^T + A\right|$. Note that the obtained result does not contradict Example 2.15.

Example 2.16: Consider a fifth-order distillation column example with the system matrix given by (Petkov et al., 1986)

$$A = \begin{bmatrix} -0.1094 & 0.0628 & 0 & 0 & 0 \\ 1.3060 & -2.1320 & 0.9807 & 0 & 0 \\ 0 & 1.5950 & -3.1490 & 1.5470 & 0 \\ 0 & 0.0355 & 2.6320 & -4.2570 & 1.8550 \\ 0 & 0.0023 & 0 & 0.1636 & -0.1625 \end{bmatrix}$$

With $Q = I$ we obtain $det(P) = |P| = 0.0544$. Theorem 2.14, formulas (2.47) and (2.48), and Theorem 2.15, respectively, imply the following lower bound for the determinant of the solution matrix P

$$Theorem\ 2.14 \Rightarrow det(P) \geq 1.8713$$

$$Formula\ (2.47) \Rightarrow det(P) \geq 0.011$$

$$Formula(2.48) \Rightarrow det(P) \geq 3.28 \times 10^{-6}$$

$$Theorem\ 2.15 \Rightarrow det(P) \geq 1.8713$$

Note that the formula (2.49) is not applicable in this case since $\lambda_1\left(A + A^T\right) > 0$. However, for the gas absorber problem from Example 2.14 this condition is satisfied so that the upper bound from (2.49) gives a remarkable result, that is

$$0.0531 \leq det(P) = 0.0544 \leq 0.0549$$

where the lower bound is obtained also by using Komaroff's result from Theorem 2.15.

$$\Delta$$

2.3 Numerical Solutions

Numerical solution of the Lyapunov matrix equation is perhaps the most important among all of the categories discussed in this book. The importance of such a solution is manifested due to practical use of the solution in different engineering and mathematical applications. Therefore, finding an efficient numerical solution has received a great deal of attention from researchers. The development and evolution of such numerical techniques are the subjects of this section.

In principle one can obtain the solution of the continuous-time algebraic Lyapunov equation (2.1) by using the skew-symmetric matrix approach and solving $0.5n(n-1)$ linear algebraic equations as explained in Section 2.1.2 by employing any of the conventional techniques for solving the system of linear algebraic equations. The solution of (2.1) can be also found by transforming the system matrix into some canonical forms, like the Jordan form (Ma, 1966) or companion form (Smith, 1966), and then using the explicit formulae described earlier. However, such approaches require large memory and computer processing time. The processing time is of the order of $n^6\mu$ or $n^5\mu$, where μ is the time for one multiplication or division. This is a very high figure for large n and becomes impractical even for moderate n, for example, $n > 10$. In (Jameson, 1968) a method is described that requires $O(n^4)$ multiplications, but still this is high and there is a need to cut down such large processing time and storage requirement. For most of these direct and transformation approaches and their comparison, the reader is referred to (Rothschild and Jameson, 1970) and (Hagander, 1972).

We first present several classic algorithms for numerical solution of the algebraic Lyapunov equation and then in an independent section we consider the most important one (Bartels and Stewart, 1972) and comment on its relationship to the algorithms of (Golub et al., 1979) and (Hammarling, 1982). All the algorithms described next require $O(n^2)$ memory locations and $O(n^3)$ multiplications.

Before describing classic numerical algorithms, it is worth mentioning here that iterative techniques and parallel algorithms for solving large scale algebraic Lyapunov equations have been developed recently. Due to their complexity and importance they will be presented in Chapter 7.

The first method to be described here calculates the numerical integration of the solution given by

$$P = \int_0^\infty e^{A^T t} Q e^{At} dt \tag{2.50}$$

The method used for this integration is the Crank-Nicolson method (Davison, 1967). The algorithm is presented in (Davison and Man, 1968). It consists of the following two steps.

Algorithm 2.4: (Davison and Man, 1968)

1. Initialize $P_0 = hQ$ for some small $h > 0$.
2. For $k = 0, 1, 2, \ldots$ do

$$C = (I - \frac{h}{2}A + \frac{h^2}{12}A^2)^{-1}(I + \frac{h}{2}A + \frac{h^2}{12}A^2)$$

$$P_{k+1} = (C^T)^{2^k} P_k C^{2^k} + P_k$$

$$\Delta$$

This algorithm is stable and converges for all small values of h. A typical value of h is

$$h = \frac{1}{200|\lambda_{max}(A)|}$$

where $\lambda_{max}(A)$ is the eigenvalue of A with maximal real part. This algorithm requires $4n^2$ words of memory and $(2.5k + 4)n^3$ multiplications. k is the number of iterations required for desired accuracy. Typically k varies from $5 < k < 20$ for 6 digit accuracy. A higher order of accuracy iterative scheme for the above algorithm is developed in (Man, 1971).

An improved method of obtaining P_0 for Algorithm 2.4 can be employed (Hagander, 1972). This method uses the following iterations

$$T_0 = hQ, \quad T_{k+1} = \frac{h}{k+1}\left[T_k A + (T_k A)^T\right]$$
$$P_0^{(0)} = T_0, \quad P_0^{(k+1)} = P_0^{(k)} + T_{k+1} \tag{2.51}$$

which are obtained by series expansion of

$$P_0 = \int_0^h e^{A^T t} Q e^{At} dt$$

An interesting and important algorithm is described in (Smith, 1968). It is very efficient for solving high-order Lyapunov algebraic equations. Smith's algorithm will be presented in detail in Chapter 7, where we study iterative methods for solving large Lyapunov equations (Wachspress, 1988; Starke, 1992). This algorithm requires $2.5n^2$ words of memory and $2.5n^3(k + 1)$ multiplications, where k is the number of iterations required for desired accuracy.

Preview and comparison of the methods presented so far and other methods for numerical solution of the algebraic Lyapunov equation known before 1972 are given in (Pace and Barnett, 1972).

At the end of this subsection, it is worth mentioning another simple and fast algorithm as described in (Hoskins et al., 1977), which is applicable when A has real spectrum (Barraud, 1979). Moreover, the algorithm is extended in (Barraud, 1979) so that it becomes applicable for any spectrum of A. The algorithm is as follows.

Algorithm 2.5: (Hoskins et al., 1977)

1. Estimate the spectrum of A. Let it be in the interval $[b, B]$.
2. Evaluate

$$\alpha = \frac{2b}{\left(b + \sqrt{bB}\right)^2}, \quad \beta = bB\alpha, \quad \epsilon = \left(\frac{b - \sqrt{bB}}{b + \sqrt{bB}}\right)^2$$

3. Compute

$$A_m = \alpha A_{m-1} + \beta A_{m-1}^{-1}, \quad A_0 = A$$
$$P_m = \alpha P_{m-1} + \beta A_{m-1}^{-T} P_{m-1} A_{m-1}^{-1}, \quad P_0 = Q$$

4. Compute new bounds for the spectrum of A_m as $[b, B] = [1 - \epsilon, 1 + \epsilon]$
5. Repeat steps 2–4 until A_m is close to the identity matrix I. Suppose that for $m = k, A_m$ is close to I.
6. The sought solution is given by $P = -0.5P_k$.

$\triangle$

This algorithm has the property that it is stable when the spectrum of A is real and "drastically better than the Newton algorithm" (obtained for $\alpha = \beta = 0.5$), (Barraud, 1979). Typical number of iterations required for convergence of A_m to I is five. It requires approximately $17n^3$ multiplications and $4n^2$ words of memory.

If the spectrum of A is complex, one can use the overdetermined values of b and B as (Barraud, 1979)

$$b = \frac{1}{||A^{-1}||} , \qquad B = ||A||$$

where $||A||$ is the spectral norm of A and calculate the corrected coefficients α and β as

$$\alpha = \frac{1}{2\sqrt{bB}}, \quad \beta = \frac{1}{2}\sqrt{bB} \tag{2.52}$$

and then execute Algorithm 2.5 with corrected coefficients (2.52). More over, it has been indicated in (Barraud, 1979) that the algorithm of (Hoskins et al., 1977) belongs to the class of the so-called matrix sign algorithms.

For the complete understanding of the matrix sign methods the interested reader is referred to (Denman and Beavers, 1976; Roberts, 1980; Blazer, 1980). The matrix sign method for numerical solution of the algebraic Lyapunov equation was originally developed in (Roberts, 1971) in an internal report, which was reprinted in (Roberts, 1980). Independently, by using a little bit different reasoning, the matrix sign method for solving numerically the continuous-time algebraic Lyapunov equation was derived by (Beavers and Denman, 1975). A very comprehensive overview of the matrix sign method techniques for solving algebraic equations appearing in systems and control theory can be found in (Denman and Beavers, 1976). In (Balzer, 1980) the accelerated matrix sign techniques obtained by using proper scaling in each iteration step are considered.

It should be pointed out that even though the matrix sign methods are considered as very powerful tools for solving algebraic equations, the rigorous numerical efficiency comparison study between the matrix sign method for solving the algebraic Lyapunov equation and the Bartels-Stewart algorithm (to be presented in the next subsection) is not yet available in the literature.

2.3.1 Bartels and Stewart Algorithm

Next we describe the most widely used algorithm obtained by (Bartels and Stewart, 1972). Due to its importance and use, this algorithm will be described in a little bit of detail. The algorithm as described in (Bartels and Stewart, 1972) computes the solution of the general Sylvester equation

$$XP + PY + Q = 0 \qquad (2.53)$$

whose special case is the Lyapunov equation, that is, when $X = A^T$ and $Y = A$. The description to be given here applies only to the Lyapunov equation. Note that *this algorithm does not require that A is an asymptotically stable matrix.*

The algorithm is based on the Schur reduction to triangular form by orthogonal similarity transformation. Let U be orthogonal matrix such that

$$\tilde{A} = U^T A U = \begin{bmatrix} \tilde{A}_{11} & \tilde{A}_{12} & \ldots & \tilde{A}_{1p} \\ 0 & \tilde{A}_{22} & \ldots & \tilde{A}_{2p} \\ \vdots & \vdots & \ddots & \vdots \\ 0 & 0 & \ldots & \tilde{A}_{pp} \end{bmatrix} \qquad (2.54)$$

is in upper real Schur form. Similarly let

$$\tilde{Q} = -U^T Q U = \begin{bmatrix} \tilde{Q}_{11} & \ldots & \tilde{Q}_{p1}^T \\ \vdots & \ddots & \vdots \\ \tilde{Q}_{p1} & \ldots & \tilde{Q}_{pp} \end{bmatrix} \qquad (2.55)$$

and

$$\tilde{P} = U^T P U = \begin{bmatrix} \tilde{P}_{11} & \ldots & \tilde{P}_{p1}^T \\ \vdots & \ddots & \vdots \\ \tilde{P}_{p1} & \ldots & \tilde{P}_{pp} \end{bmatrix} \qquad (2.56)$$

then equation (2.1) is equivalent to

$$\tilde{A}^T \tilde{P} + \tilde{P} \tilde{A} = \tilde{Q}$$

Note that since P and Q are symmetric matrices, $\widetilde{P}$ and $\widetilde{Q}$ are also symmetric matrices. If the partitions of $\widetilde{A}$, $\widetilde{P}$, and $\widetilde{Q}$ are compatible, then the following equations can be formed. For $l = 1, 2, \ldots, p$

$$\widetilde{A}_{kk}^T \widetilde{P}_{kl} + \widetilde{P}_{kl} \widetilde{A}_{ll} = \widetilde{Q}_{kl} - \sum_{j=1}^{k-1} \widetilde{A}_{jk}^T \widetilde{P}_{jl} - \sum_{i=1}^{l-1} \widetilde{P}_{ki} \widetilde{A}_{il} \qquad (2.57)$$

$$k = l, l+1, ..., p$$

The solution of these equations still requires the solution of the algebraic Lyapunov equations. However, since matrices $\widetilde{A}_{kk}$ and $\widetilde{A}_{ll}$ are of order at most two, hence the solutions of (2.57) can be obtained by solving a linear system of order at most four. Moreover, when $k = l$, one has to calculate at most three linear equations for three distinct elements of $\widetilde{P}_{kk}$, $k = 1, 2, ..., p$. In some cases only solutions of scalar algebraic Lyapunov equations are required, see Example 2.17. The system of linear equations is solved by the Crout reduction in the original paper (Bartels and Stewart, 1972).

The reduction of A to upper real Schur form is carried out by first reducing A to upper Hessenberg form by Householder's method and then applying the QR algorithm (Wilkinson, 1965) to obtain the desired upper Schur form (Martin et al., 1970). The product of the transformations used in the reduction process forms the matrix U.

The Schur form can be easily obtained by using a MATLAB statement [U,Atilde]=schur(A), where schur stands for the MATLAB function that computes both the Schur form $\widetilde{A}$ and the corresponding transformation matrix U. The complete algorithm is as follows.

Algorithm 2.6: (Bartels and Stewart, 1972)

1. Transform A to upper real Schur form $\widetilde{A} = U^T A U$ as in equation (2.54).
2. Construct $\widetilde{Q} = -U^T Q U$ as in equation (2.55).
3. For $l = 1, 2, \ldots, p$ and $k = l, l+1, \ldots, p$, solve (2.57) for matrix $\widetilde{P}$.
4. Construct the required solution by $P = U \widetilde{P} U^T$.

$\triangle$

The number of multiplications for this algorithm depends on the criterion adopted in iterative QR algorithm for determining whether a subdiagonal element of upper Hessenberg matrix is negligible. If σ is the average number of QR steps required to make a subdiagonal element negligible, then the number of multiplications required to solve (2.1) by this algorithm is estimated by $4n^3(\sigma + 1)$. According to (Wachspress, 1992) the Bartels-Stewart algorithm requires $15n^3$ flops. The storage requirement is $2.5n^2 + n/2$ locations to hold A, U, and Q. A complete FORTRAN code for this algorithm is given in (Bartels and Stewart, 1972).

Example 2.17: For the matrices A and Q of the industrial reactor presented in Example 2.7, we obtain the following $\tilde{A}$ and U after performing the Schur reduction

$$\tilde{A} = \begin{bmatrix} -128.5021 & 89.2591 & 40.2984 & -223.6790 & 28.0844 \\ 0 & -74.4361 & -1.9334 & 18.6046 & 8.6728 \\ 0 & 0 & -2.9376 & 14.2652 & -0.3919 \\ 0 & 0 & 0 & -82.1951 & -49.7840 \\ 0 & 0 & 0 & 0 & -8.8190 \end{bmatrix}$$

$$U = \begin{bmatrix} 0.0781 & 0.4902 & 0.8607 & 0.1111 & 0.0203 \\ -0.0116 & 0.0267 & 0.0262 & -0.1310 & -0.9906 \\ -0.3228 & -0.7946 & 0.4980 & -0.1272 & 0.0124 \\ -0.9374 & 0.2912 & -0.1016 & 0.1616 & -0.0052 \\ 0.1040 & -0.2067 & -0.0129 & 0.9634 & -0.1345 \end{bmatrix}$$

Since the matrix $\tilde{A}$ is pure upper triangular, equation (2.57) requires only solution of scalar algebraic Lyapunov equations. Also since $Q = I_5$, we have from (2.55) $\tilde{Q} = -Q = -I$. Calculating (2.57) we obtain the matrix $\tilde{P}$. For example, for $l = 1, k = 1$, equation (2.57) implies

$$2 \times (-128.5021)\tilde{P}_{11} = -1 \Rightarrow \tilde{P}_{11} = 0.003891$$

$l = 1, k = 2$ produces

$$(-128.5021 - 74.4361)\tilde{P}_{11} = \tilde{Q}_{21} - \tilde{P}_{11}(89.2591) \Rightarrow \tilde{P}_{21} = 0.001711$$

$l = 1, k = 3$ yields to

$$\left(\tilde{A}_{33} + \tilde{A}_{11}\right)\tilde{P}_{31} = \tilde{Q}_{31} - \tilde{A}_{13}\tilde{P}_{11} - \tilde{A}_{23}\tilde{P}_{21} \Rightarrow \tilde{P}_{31} = 0.001168$$

and so on, leads to

$$\tilde{P} = \begin{bmatrix} 0.0039 & 0.0017 & 0.0012 & -0.0039 & 0.0023 \\ 0.0017 & 0.0088 & 0.0020 & -0.0034 & 0.0060 \\ 0.0012 & 0.0020 & 0.1849 & 0.0266 & -0.1075 \\ -0.0039 & -0.0034 & 0.0266 & 0.0205 & -0.0342 \\ 0.0023 & 0.0060 & -0.1075 & -0.0342 & 0.2678 \end{bmatrix}$$

Finally, the required matrix P is calculated from $P = U\tilde{P}U^T$, which turns out to be exactly the same as in Example 2.7.

$\triangle$

If one is interested in finding the solution of the general Sylvester equation (2.53), then a modified version of Bartels and Stewart's algorithm, so-called the Hessenberg-Schur method, can be also used as described in (Golub et al. 1979). In such a modified algorithm, the matrix X as in equation (2.53), is reduced only to the Hessenberg form instead of the Schur form, while matrix Y is still reduced to the Schur form. The algorithm is faster than Bartels and Stewart's algorithm, especially when the dimension of X is larger than the dimension of Y, but the storage requirement is greater.

Recently, the algorithm of (Hammarling, 1982) has become very popular since his main features are comparable to those of (Bartels and Stewart, 1972). As a matter of fact the Hammarling algorithm is also based on the Schur decomposition and represents a variant of the Bartels-Stewart algorithm.

Table 2.2 presents a comparison of the storage requirements and the number of multiplications needed for each of the four algorithms considered in Section 2.3. In order to determine the efficiency of any of these algorithms, one has to consider other conditions also, for example stability of matrix A. In Table 2.2, n is the size of matrix P, k is the number of iterations required for desired accuracy, and σ is the number of QR steps in the Bartels-Stewart algorithm.

Finally, let us point out that the matrix sign method for numerical solution of the generalized algebraic Lyapunov equation of the form

$$A^T P E + E^T P A + Q \qquad (2.58)$$

with AE^{-1} stable has been derived in (Aliev and Larin, 1993).

Algorithm	Storage	Number of Multiplications
Davison and Man, 1968	$4n^2$	$(2.5k + 4)n^3$
Smith, 1968	$2.5n^2$	$2.5(k + 1)n^3$
Hoskins, 1977	$4n^2$	$17n^3$
Bartels and Stewart, 1972	$2.5n^2 + 0.5n$	$4n^3(\sigma + 1)$

Table 2.2: Comparison of four algorithms

2.4 Summary

In this chapter solutions for the continuous algebraic Lyapunov equation are presented. The analytical solutions mainly involve expanding the algebraic matrix Lyapunov equation into a system of linear equations, which can be solved by using the standard techniques. The Kronecker product method is very simple and elegant, but it yields to a very high order system (a system of order n^2), which for large n is difficult to solve. The methods which develop lower order systems take the advantage of the symmetry of the solution matrix P to yield $n(n + 1)/2$ linear equations. Two algorithms are presented in this respect. The use of a skew-symmetric matrix can even further reduce the number of linear equations to $n(n - 1)/2$. The transformation approaches involve transforming the Lyapunov matrix equation into the one where the coefficient matrix has some canonical form like diagonal or Jordan form. The transformed equation is then easier to solve because of the special structure of the coefficient matrix. Solutions of the Lyapunov matrix equations with some special structures for the coefficient matrix are also discussed. This includes the cases when the coefficient matrix is in Schwarz form, companion form, and Jordan form. Several examples are also presented in order to demonstrate the solution procedures.

Approximate solutions and bounds on different parameters of the solution matrix are also presented. These parameters include eigenvalues, trace, and determinant. The bounds obtained can be used to examine the behavior of the underlying system without solving the Lyapunov equation. Some examples are presented to show the tightness of the bounds. Comparison is made wherever possible. It is important to point out that the issue with bounds of the solution attributes for the algebraic Lyapunov equation remains still an open research area.

Numerical solutions are very important in order to compute the solution on the digital computers. Therefore, a coverage of all major numerical algorithms is given. Instead of going through the detailed theory underlying these algorithms, they are described from the practical point of view. Four algorithms are presented, among which all but Algorithm 2.6 are iterative algorithms. Algorithm 2.6 (Bartels-Stewart) is the most widely used algorithm, and it employs Schur decomposition and then builds simple linear equations which can be solved sequentially. These algorithms are designed to reduce the number of operations and the storage requirements. At the end a comparison is given for all of the four algorithms.

2.5 References

1. Aliev, F. and V. Larin, "Generalized Lyapunov equation and factorization of matrix polynomials," *Systems & Control Letters*, vol.21, 485–491, 1993.

2. Arkun, Y. and S. Ramakrishnan, "Bounds of the optimal quadratic cost of structure constrained regulators," *IEEE Trans. Automatic Control*, vol.28, 924-927, 1983.

3. Balzer, L., "Accelerated convergence of the matrix sign function method of solving Lyapunov, Riccati and other matrix equations," *Int. J. Control*, vol.32, 1057–1078, 1980.

4. Barnett, S., "Simplification of Lyapunov matrix equation $A^T P A - P = -Q$," *IEEE Trans. Automatic Control*, vol.19, 446–447, 1974.

5. Barnett, S. and T. Man, "Comments on 'A theorem on the Lyapunov matrix equation'," *IEEE Trans. Automatic Control*, vol.15, 279-280, 1970.

6. Barnett, S. and C. Storey, "Stability analysis of constant linear systems by Liapunov second method," *Electronics Letters*, vol.2, 165–166, 1966a.

7. Barnett, S. and C. Storey, "Solution of the Liapunov matrix equation," *Electronics Letters*, vol.2, 466–467, 1966b.

8. Barnett, S. and C. Storey, "The Liapunov matrix equation and Schwarz's form," *IEEE Trans. Automatic Control*, vol.12, 117–118, 1967a.

9. Barnett, S. and C. Storey, "On the general functional matrix for a linear system," *IEEE Trans. Automatic Control*, vol.12, 436–438, 1967b.

10. Barnett, S. and C. Storey, "Remarks on numerical solution of the Liapunov matrix equation," *Electronics Letters*, vol.3, 416–417, 1967c.

11. Barnett, S. and C. Storey, "Further remarks on the Lyapunov equation and Schwarz's form," *IEEE Trans. Automatic Control*, vol.13, 204-205, 1968.

12. Barraud, A. "Comments on 'The numerical solution of $A^T Q + QA = -C$'," *IEEE Trans. Automatic Control*, vol.24, 671–672, 1979.

13. Bartels, R. and G. Stewart, "Algorithm 432, solution of the matrix equation $AX + XB = C$," *Comm. Ass. Computer Machinery*, vol.15, 820–826, 1972.

14. Beavers, A. and E. Denman, "A new solution method for the Lyapunov matrix equation," *SIAM J. Appl. Math.*, vol.29, 416–421, 1975.

15. Bellman, R., "Kronecker products and the second method of Lyapunov," *Mathematische Nachrichten*, 17–19, Berlin, Akademie-Verlag, 1959.

16. Bialas, S., "On the Lyapunov matrix equation," *IEEE Trans. Automatic Control*, vol.25, 813-814, 1980.

17. Bingulac, S., "An alternate approach to expanding $PA + A^T P = -Q$," *IEEE Trans. Automatic Control*, vol.15, 135–136, 1970.

18. Butchart, R., "Explicit solution to the Fokker Plank equation for an ordinary differential equation," *Int. J. Control*, vol.1, 201–208, 1965.

19. Chen, C., *Linear System Theory and Design*, Holt, Rinehart and Winston, New York, 1984.

20. Chen, C. and H. Chu, "A matrix for evaluation Schwarz's form," *IEEE Trans. Automatic Control*, vol.11, 303-305, 1966.

21. Chen, C. and L. Shieh, "A note on expanding $PA + A^T P = -Q$," *IEEE Trans. Automatic Control*, vol.13, 122–123, 1968.

22. Curran, P., "Lyapunov's matrix equation with system matrix in companion form," *Int. J. Control*, vol.57, 1509–1516, 1993.

23. Davison, E., "A high order Crank–Nicolson technique for solving differential equations," *Computer J.*, vol.10, 195–197, 1967.

24. Davison, E. and F. Man, "The numerical solution of $A^T Q + QA = -C$," *IEEE Trans. Automatic Control*, vol.13, 448–449, 1968.

25. Denman, E. and A. Beavers, "The matrix sign function and computations in systems," *Appl. Math. and Computation*, vol.2, 63–94, 1976.

26. De Vlieger, J., H. Verbruggen, and P. Bruijn, "A time-optimal control algorithm for digital computer control," *Automatica*, vol.18, 239–244, 1982.

27. Fan, K., "Minimum properties and inequalities for the eigenvalues of completely continuous operators," *Proc. National Academy of Science, USA*, vol.37, 760–766, 1951.

28. Fan, K., "A minimum property of the eigenvalues of a Hermitian transformation," *Amer. Math. Monthly*, vol.60, 48–50, 1953.

29. Fan, K. and A. Hoffman, "Some metric inequalities in the space of matrices," *Proc. Amer. Math. Soc.*, vol.6, 111–116, 1955.

30. Fahmy, M. and A. Hanafy, "Further correction to: Comments on 'On the Lyapunov matrix equation'," *IEEE Trans. Automatic Control*, vol.26, 619, 1981.

31. Gantmacher, F., *The Theory of Matrices*, vol.1 and 2, Chelsea, New York, 1959.

32. Garloff, J., "Bounds for eigenvalues of the solution of the discrete Riccati and Lyapunov equations and the continuous Lyapunov equation," *Int. J. Control*, vol.43, 423-431, 1986.

33. Golub, G., S. Nash, and C. Loan, "A Hessenberg-Schur method for the problem $AX + XB = C$," *IEEE Trans. Automatic Control*, vol.24, 909–913, 1979.

34. Guidorzi, R., "Transformation approach to the solution of matrix equation $A^T X + XB = P$," *IEEE Trans. Automatic Control*, vol.17, 377-379, 1972.

35. Hagander, P., "Numerical solution $A^T S + SA + Q = 0$," *Infor. Science*, vol.4, 35-50, 1972.

36. Hammarling, S., "Numerical solution of the stable, non-negative definite Lyapunov equation," *IMA J. Numerical Analysis*, vol.2, 303-323, 1982.

37. Hmamed, A., "Differential and difference Lyapunov equations: simultaneous eigenvalue bounds," *Int. J. Systems Science*, vol.21, 1335-1344, 1990.

38. Hodel, A. "The recent application of the Lyapunov equation in control theory," in *Iterative Methods in Linear Algebra*, 217–227, R. Beauwens and P. deGroen, Eds., North-Holland, Amsterdam, 1992.

39. Horn, A., "On the singular values of a product of completely continuous operators," *Proc. National Academy of Science, USA*, vol.36, 374–375, 1950.

40. Hoskins, W., D. Meek, and D. Walton, "The numerical solution of $A^T Q + QA = -C$," *IEEE Trans. Automatic Control*, vol.22, 882-883, 1977.

41. Householder, A., *The Theory of Matrices in Numerical Analysis*, Blaisdell, 1964.

42. Jacyno, Z., "Explicit direct solution of Lyapunov matrix equation," *J. Franklin Institute*, vol.326, 793-801, 1989.

43. Jameson, A., "Solution of the equation $AX + XB = C$ by inversion of $M \times M$ or $N \times N$ matrix," *SIAM J. Appl. Math.*, vol.16, 1020-1023, 1968.

44. Kalman, R. and J. Bertram, "Control system analysis and design via the 'second method' of Lyapunov: I Continuous-time systems," *Trans. ASME J. Basic Engineering, Series D.*, vol.82, 371-393, 1960.

45. Karanam, V., "Lower bounds on the solution of Lyapunov matrix and algebraic Riccati equation," *IEEE Trans. Automatic Control*, vol.26, 1288-1290, 1981.

46. Karanam, V., "Eigenvalue bounds for algebraic Riccati and Lyapunov equations," *IEEE Trans. Automatic Control*, vol.27, 461-463, 1982.

47. Karanam, V., "A note on eigenvalue bounds in algebraic Riccati equation," *IEEE Trans. Automatic Control*, vol.28, 109–111, 1983.

48. Karanam, V., "Correction to 'Eigenvalue bounds for algebraic Riccati and Lyapunov equations," *IEEE Trans. Automatic Control*, vol.31, 92, 1986.

49. Kokotovic, P., "Feedback design of large scale linear systems," in *Feedback Systems*, by J. Cruz, McGraw-Hill, New York, 1972.

50. Kokotovic, P., J. Allemong, J. Winkelman, and J. Chow, "Singular perturbations and iterative separation of the time scales," *Automatica*, vol.16, 23–33, 1980.

51. Komaroff, N., "Simultaneous eigenvalue lower bounds for Lyapunov matrix equation," *IEEE Trans. Automatic Control*, vol.33, 126-128, 1988.

52. Komaroff, N., "Upper bounds for the eigenvalues of the solution of the Lyapunov matrix equation," *IEEE Trans. Automatic Control*, vol.35, 737-739, 1990.

53. Komaroff, N., "Upper summation and product bounds for solution eigenvalues of the Lyapunov matrix equation," *IEEE Trans. Automatic Control*, vol.37, 1040–1042, 1992.

54. Kwon, W. and A. Pearson, "A note on algebraic matrix Riccati equation," *IEEE Trans. Automatic Control*, vol.22, 143-144, 1977.

55. Kwon, B., M. Youn, and Z. Bien, "On bounds of the Riccati and Lyapunov matrix equations," *IEEE Trans. Automatic Control*, vol.30, 1134-1135, 1985.

56. Kwon, B. and M. Youn, "Comments on 'On some bounds in the algebraic Riccati and Lyapunov equations'," *IEEE Trans. Automatic Control*, vol.31, 591, 1986.

57. Lancaster, P. "Explicit solutions of linear matrix equations," *SIAM Review*, vol.12, 544-566, 1970.

58. Lancaster, P. and M. Tismenetsky, *Theory of Matrices*, Academic Press, New York, 1985.

59. Litkouhi, B., *Sampled-Data Control of Systems with Slow and Fast Modes*, Ph. D. Dissertation, Michigan State University, 1983.

60. Lu, C., "Solution of the matrix equation $AX + XB = C$," *Electronics Letters*, vol.7, 185–186, 1971.

61. Ma, E., "A finite series solution of the matrix equation $AX - XB = C$," *SIAM J. Appl. Math.*, vol.14, 490–495, 1966.

62. MacFarlane, A., "The calculation of functionals of the time and frequency response of a linear constant coefficient dynamical system," *Quart. Journ. Mech. and Applied Math.*, Pt.2, vol.16, 259–271, 1963.

63. Man, F., "A theorem on the Lyapunov matrix equation," *IEEE Trans. Automatic Control*, vol.14, 306, 1969.

64. Man, F., "A high order method of solution for the Lyapunov matrix equation," *Computer J.*, vol.14, 291–292, 1971.

65. Martin, R., G. Peters, and J. Wilkinson, "The QR algorithm for real Hessenberg matrices," *Numer. Math.*, vol.14, 219–231, 1970.

66. Mitrinovic, D., *Analytic Inequalities*, Springer Verlag, New York, 1970.

67. Montemayor, J. and B. Womack, "Comments on 'On the Lyapunov matrix equation'," *IEEE Trans. Automatic Control*, vol.20, 814-815, 1975.

68. Mori, T. "On some bounds in algebraic Riccati and Lyapunov equations," *IEEE Trans. Automatic Control*, vol.30, 162-164, 1985.

69. Mori, T., N. Fukuma, and M. Kuwahara, "A note on the Lyapunov matrix equation," *IEEE Trans. Automatic Control*, vol.26, 941-942, 1981.

70. Mori, T., N. Fukuma, and M. Kuwahara, "Explicit solution and eigenvalue bounds in the Lyapunov matrix equation," *IEEE Trans. Automatic Control*, vol.31, 656-658, 1986a.

71. Pace, I. and S. Barnett, "Comparison of numerical methods for solving Liapunov matrix equations," *Int. J. Control*, vol.15, 907-915, 1972.

72. Patel, R. and M. Toda, "On norm bounds for algebraic Riccati and Lyapunov equations," *IEEE Trans. Automatic Control*, vol.23, 87-88, 1978.

73. Peng, T., "A note on the Lyapunov matrix equation," *IEEE Trans. Automatic Control*, vol.17, 565, 1972.

74. Petkov, P., N. Christov, and M. Konstantinov, "A computational algorithm for pole placement assignment of linear multi-input systems," *IEEE Trans. Automatic Control*, vol.31, 1044–1047, 1986.

75. Power, H., "Solution of Lyapunov matrix equation for continuous systems via Schwarz and Routh canonical forms," *Electronics Letters*, vol.3, 81–82, 1967a.

76. Power, H., "Further comments on the Lyapunov matrix equation," *Electronics Letters*, vol.3, 153–154, 1967b.

77. Power, H., "Solution of the Lyapunov matrix equation with a diagonal input matrix, obtained without matrix inversion," *Electronics Letters*, vol.3, 325–326, 1967c.

78. Power, H., "Solution of the Lyapunov matrix equation with a diagonal input matrix, obtained without matrix inversion," *Electronics Letters*, vol.5, 135–136, 1969.

79. Roberts, J., *Report CUED/B — Control/TR13*, Engineering Department, Cambridge University, 1971.

80. Roberts, J., "Linear model reduction and solution of the algebraic Riccati equation by use of the sign function," *Int. J. Control*, vol.32, 677–687, 1980.

81. Rothschild, D., and A. Jameson, "Comparison of four numerical algorithms for solving the Liapunov matrix equation," *Int. J. Control*, vol.11, 181-198, 1970.

82. Rutherford, D., "On the solution of the matrix equation $AX + XB = C$," *Nederl. Akad. Wetensch. Proc. Ser.A.*, vol.35, 53–59, 1932.

83. Schwarz, H., "A method for determining stability of matrix differential equations," *Z. angew. Math. Phys.* vol.7, 473–500, 1956.

84. Shapiro, E., *Suboptimal Design of Minimal-Time Feedback Regulators for Linear, Time-Invariant Plants*, D. Sc. Dissertation, Columbia University, New York, 1972.

85. Shapiro, E., "On the Lyapunov matrix equation," *IEEE Trans. Automatic Control*, vol.19, 594–596, 1974.

86. Shen, X. and A. Zelentsovsky, "Comments on 'Explicit direct solution of the Lyapunov matrix equation'," *J. Franklin Institute*, vol.328, 519–522, 1991.

87. Smith, R., "Bounds for Liapunov quadratic forms," *J. Math. Anal. and Appl.*, vol.12, 425–435, 1965.

88. Smith, R., "Matrix calculations for Liapunov quadratic forms," *J. Diff. Equations*, vol.2, 208–217, 1966.

89. Smith, R., "Matrix equation $XA + BX = C$," *SIAM J. Appl. Math.*, vol.16, 198–201, 1968.

90. Starke, G., "SOR-like methods for Lyapunov matrix equations," in *Iterative Methods in Linear Algebra*, 233–240, R. Beauwens and P. deGroen, Eds., North-Holland, Amsterdam, 1992.

91. Sreeram and Agathoklis, "Solution of Lyapunov equation with system matrix in companion form," *Proc. IEE, Part D.*, vol.138, 529–534, 1991.

92. Troch, I., "Improved bounds for eigenvalues of solution of Lyapunov equations," *IEEE Trans. Automatic Control*, vol.32, 744-747, 1987.

93. Xiao, C., Z. Feng, and X. Shan, "On the solution of the continuous-time Lyapunov matrix equation in two canonical forms," *Proc. IEE, Part D.*, vol.139, 286–300, 1992.

94. Wachspress, E. "Iterative solution of the Lyapunov matrix equation," *Applied Math. Letters*, vol.1, 87–90, 1988.

95. Wang, S., T. Kuo, and C. Hsu, "Trace bounds on the solution of algebraic matrix Riccati and Lyapunov equations," *IEEE Trans. Automatic Control*, vol.31, 654-656, 1986.

96. Wilkinson, J., *The Algebraic Eigenvalue Problem*, Oxford University Press, Cambridge, 1965.

97. Wimmer, H., "Generalizations of theorems of Lyapunov and Stein," *Linear Algebra and Its Appl.*, vol.10, 139-146, 1975.

98. Yasuda, K. and K. Hirai, "Upper and lower bounds on solution of algebraic Riccati equation," *IEEE Trans. Automatic Control*, vol.24, 483–487, 1979.

99. Ziedan, I., "Explicit solution of the Lyapunov matrix equation," *IEEE Trans. Automatic Control*, vol.17, 379–381, 1972.

Chapter Three

Discrete Algebraic Lyapunov Equation

Similarly to the previous chapter, where we have studied the continuous algebraic Lyapunov equation, here the discrete algebraic Lyapunov equation is considered in three categories, namely, explicit solutions, bounds of solution's attributes, and numerical solutions. In the contents of explicit solutions, we present the bilinear transformation technique (Popov, 1964; Power, 1967, 1969), skew-symmetric matrix approach (Barnett, 1974), and Jordan form technique (Heinen, 1972). Several results are given for the solution bounds. According to the work of Mori and his coworkers, and Komaroff, similarly to the case of the continuous algebraic Lyapunov equation, no definite answers have been obtained yet about the best possible bounds for the solution attributes (eigenvalues, trace, determinant) of the discrete-time algebraic Lyapunov equation. Obtained bounds are mostly based on different inequalities such as Minkovski's inequality for determinants and inequalities from (Ostrowski, 1959; Weyl, 1949; Fan, 1949; Amir-Moez, 1956) so that it is hard to make definite analytical comparisons. It is the authors' opinion that the papers of (Mori et al., 1982a, 1982b), (Garloff, 1986), and (Komaroff, 1992) contain the most comprehensive results so that our presentation of the solution bounds is mostly based on their work. As far as the numerical solutions are concerned, we present the result of (Barraud,

1977), which is similar to the Bartels-Stewart algorithm for numerical solution of the continuous algebraic Lyapunov equation.

3.1 Explicit Solutions

The approaches to solve the algebraic Lyapunov matrix equation for discrete systems are similar to those for continuous systems. Like for the continuous algebraic Lyapunov equation, the procedures may involve expanding the discrete algebraic Lyapunov matrix equation into a system of linear equations, or converting it into some canonical forms through some nonsingular transformations. Among these techniques, the most commonly used method is to convert the discrete algebraic Lyapunov equation into the continuous algebraic Lyapunov equation by using the bilinear transformation, and then to use any method as described in Chapter 2 to solve that transformed continuous Lyapunov equation. In one case, the solutions of the newly obtained continuous Lyapunov equation and the original discrete Lyapunov equation are identical (Popov, 1964). In other applications of the bilinear transformation one needs to use the inverse transformation in order to obtain the solution for the discrete algebraic Lyapunov equation (Power, 1967, 1968, 1969). In a special case when the matrices A and G ($Q = GQ'G^T$) are in controllable canonical form (Kailath, 1980), it has been shown in (Bitmead and Weiss, 1979) that the solution of the discrete algebraic Lyapunov equation is the inverse of the Schur-Cohn matrix.

3.1.1 Bilinear Transformation

The first technique to be described here makes use of the bilinear transformation in order to convert the discrete Lyapunov equation given by

$$A^T P A + Q = P \tag{3.1}$$

into its continuous counterpart

$$A^T P + P A + Q = 0 \tag{3.2}$$

Such a transformation is given in (Power, 1969), in which the following transformation is used

$$B = (A - I)^{-1}(A + I)$$

or

$$A = (B + I)(B - I)^{-1}$$

This transformation converts (3.1) into an equivalent form

$$B^T P_b + P_b B + Q = 0 \qquad (3.3)$$

where

$$P = 0.5(B - I)^T P_b(B - I)$$

Equation (3.3) is the algebraic Lyapunov equation for continuous systems and can be solved by using any of the methods described in the previous chapter.

Another form of the bilinear transformation that preserves the same solution for the continuous-time and discrete-time algebraic Lyapunov equations was derived in (Popov, 1964). It can be found also in (Kailath, 1980, page 180), where the following transformation is used

$$B = (A - I)(A + I)^{-1}$$

$$C = 2(A^T + I)^{-1}Q(A + I)^{-1}$$

It can be easily verified that (3.1) has the equivalent form

$$B^T P + PB + C = 0$$

Note that in this case we do not need to use the inverse transformation, because the solution P as obtained from the above equation is the solution of the original discrete Lyapunov equation. Both of these bilinear transformations involve the same number of operations; therefore, any one can be used without preference.

The bilinear transformation for solving the discrete-time algebraic Lyapunov equation in terms of the solution of the continuous algebraic Lyapunov equation has been also considered in (Barnett and Storey, 1970). The algebraic discrete Lyapunov equation obtained by sampling the continuous-time system and its relation to the corresponding continuous-time algebraic Lyapunov equation are studied in (Troch, 1988).

Example 3.1: Given a discrete model of a steam power system (Mahmoud et al., 1986) with

$$A = \begin{bmatrix} 0.915 & 0.051 & 0.038 & 0.015 & 0.038 \\ -0.030 & 0.889 & -0.001 & 0.046 & 0.111 \\ -0.006 & 0.648 & 0.247 & 0.014 & 0.048 \\ -0.715 & -0.022 & -0.021 & 0.240 & -0.024 \\ -0.148 & -0.003 & -0.004 & 0.090 & 0.026 \end{bmatrix}$$

Choosing $Q = I_5$ and converting the discrete Lyapunov equation into the continuous Lyapunov equation by using the bilinear transformation, which preserves the solution matrix P (Popov, 1964), we obtain the following transformed matrices

$$B = \begin{bmatrix} -0.038 & 0.017 & 0.032 & 0.009 & 0.035 \\ 0.004 & -0.058 & 0.000 & 0.031 & 0.115 \\ 0.004 & 0.550 & -0.604 & -0.003 & 0.015 \\ -0.600 & 0.001 & -0.009 & -0.604 & -0.015 \\ -0.097 & 0.002 & -0.001 & 0.142 & -0.943 \end{bmatrix}$$

$$C = \begin{bmatrix} 0.724 & -0.010 & -0.017 & 0.470 & 0.081 \\ -0.010 & 0.711 & -0.441 & -0.018 & -0.058 \\ -0.017 & -0.441 & 1.286 & 0.010 & -0.011 \\ 0.470 & -0.018 & 0.010 & 1.298 & -0.124 \\ 0.081 & -0.058 & -0.011 & -0.124 & 1.894 \end{bmatrix}$$

Then the solution of the discrete Lyapunov equation is equal to the solution of the following continuous Lyapunov equation

$$B^T P + PB + C = 0$$

and is given by

$$P = \begin{bmatrix} 10.2948 & 0.2097 & 0.4801 & -0.0974 & 0.3318 \\ 0.2097 & 8.5822 & 0.2533 & 0.5866 & 0.9606 \\ 0.4801 & 0.2533 & 1.0907 & 0.0089 & 0.0393 \\ -0.0974 & 0.5866 & 0.0089 & 1.1197 & 0.0756 \\ 0.3318 & 0.9606 & 0.0393 & 0.0756 & 1.1340 \end{bmatrix}$$

Δ

3.1.2 Jordan Form Technique

Another technique which is useful for finding the solution of (3.1) for several values of the matrix Q (Heinen, 1972), converts the matrix A to its Jordan canonical form J, that is, $J = S^{-1}AS$ for some nonsingular matrix S. Defining $E = S^T PS$ and $F = S^T QS$, equation (3.1) can be written in an equivalent form

$$J^T EJ - E + F = 0 \qquad (3.4)$$

Note that E and F are symmetric matrices because P and Q are symmetric.

For the case of m distinct eigenvalues of A, the Jordan form matrix, J, has m blocks, that is,

$$J = diag\{J_1, J_2, ..., J_m\}$$

If we partition E and F according to the dimensions of these Jordan blocks, we can obtain the following equations by expanding (3.4)

$$J_i^T E_{ij} J_j - E_{ij} + F_{ij} = 0, \quad i, j = 1, 2, 3, ..., m \qquad (3.5)$$

where E_{ij} and F_{ij} are block matrices corresponding to J_i and J_j. Denoting the elements of E_{ij} and F_{ij} by $e_{rs}^{(ij)}$ and $f_{rs}^{(ij)}$, respectively, where $r = 1, 2, \ldots, r_{ij}$ and $s = 1, 2, \ldots, s_{ij}$, the elements of E_{ij} can be calculated from (3.5) through the following recursion (Heinen, 1972)

$$e_{rs}^{(ij)} = \frac{e_{r-1,s-1}^{(ij)} + \lambda_i e_{r,s-1}^{(ij)} + \lambda_j e_{r-1,s}^{(ij)} + f_{rs}^{(ij)}}{-\lambda_i \lambda_j} \qquad (3.6)$$

Once E is determined by repeated use of (3.6), one can obtain P from

$$P = (S^{-1})^T ES$$

As mentioned earlier this method is useful when one intends to find the solution for several values of Q. In such a case, the relatively difficult step of finding the Jordan canonical form of A is not required for different values of Q.

If the matrix A is in companion form (phase variable canonical form), an expansion of (3.1) can be done analogously to the continuous-time counterpart presented in Chapter 2 (Barnett and Storey, 1967). Corresponding expansion for the discrete-time algebraic Lyapunov equation is presented in the work of (Sarma and Pai, 1968).

3.1.3 Skew-Symmetric Matrix Approach

It is worth mentioning here that the use of the skew-symmetric matrix technique, like in the continuous-time case, can be helpful for a direct solution of (3.1), which reduces the number of linear equations to be solved from $0.5n(n+1)$ to $0.5n(n-1)$, (Barnett, 1974). Using the fact that any square matrix M can be written as

$$M = M_1 + M_2$$

where M_1 is a symmetric matrix defined by

$$M_1 = \frac{1}{2}(M + M^T)$$

and M_2 is a skew-symmetric matrix given by

$$M_2 = \frac{1}{2}(M - M^T), \quad M_2 = -M_2^T$$

we can define a skew-symmetric matrix S as

$$S = PA - A^T P \tag{3.7}$$

Adding equations (3.1) and (3.7) is producing

$$(A^T + I)P(A - I) = S - Q$$

so that the required solution P can be expressed as

$$P = (A^T + I)^{-1}(S - Q)(A - I)^{-1} \tag{3.8}$$

The skew-symmetric matrix S can be obtained, after some simple algebra, from the following equation

$$A^T S A - S = A^T Q - Q A \tag{3.9}$$

Note that since both S and the right-hand side of (3.9) are skew-symmetric matrices equation (3.9) represents only $0.5n(n-1)$ linear equations and the same number of unknowns. Thus, the number of linear algebraic equations to be solved is reduced by n. However, one needs to perform inversions of two $n \times n$ matrices introduced in (3.8). Finally, the procedure is applicable under the assumption that the system matrix A has no eigenvalues at ± 1.

3.2 Bounds of Solution's Attributes

Following the same setup as with the continuous algebraic Lyapunov equation, this section describes the bounds on different parameters of the solution matrix P of the discrete algebraic Lyapunov equation (3.1). These bounds are useful in determining the various properties of the solution. Most of these bounds are obtained as an extension of the results known for the continuous algebraic Lyapunov equation.

3.2.1 Eigenvalue Bounds

The first result on a bound for the *spectral radius* of P is obtained in (Kwon and Pearson, 1977) as a by product of the corresponding bound for the algebraic discrete Riccati equation. Under assumptions that the system matrix A is invertible, $Q > 0$, and $QA^{-1} + A^{-T}Q$ is nonsingular, it holds

$$\|P\| = \max_i [\alpha_i(P)] \geq \max_i \frac{|\lambda_i(QA^{-1} + A^{-T}Q)|}{2\|A^{-1} - A\|} \tag{3.10}$$

where α_i and λ_i denote the eigenvalues of the corresponding matrices. Also, as a corollary of the work on the bounds of the algebraic discrete Riccati equation from (Patel and Toda, 1978), we have the *spectral radius bound* that is easier to obtain than the one from (Kwon and Pearson, 1977). This bound is only valid under the asymptotic stability assumption of the system matrix A and is given by

$$\|P\| \geq \frac{tr(Q)}{n - \sum\limits_{i=1}^{n} |\lambda_i(A)|^2} \tag{3.11}$$

Example 3.2: Consider a discrete model of a chemical plant (Gomathi, 1980) represented by its system matrix as

$$A = 10^{-2} \begin{bmatrix} 95.407 & 1.9643 & 0.3597 & 0.0673 & 0.0190 \\ 40.849 & 41.317 & 16.084 & 4.4679 & 1.1971 \\ 12.217 & 26.326 & 36.149 & 15.930 & 12.383 \\ 4.1118 & 12.858 & 27.209 & 21.442 & 40.976 \\ 0.1305 & 0.5808 & 1.8750 & 3.6162 & 94.280 \end{bmatrix}$$

Using $Q = I_5$, we get for the solution of the discrete algebraic Lyapunov equation (3.1) the exact value for its spectral radius as $\|P\| = 198.6889$. The lower bounds obtained from (3.10) and (3.11) produce, respectively, $\|P\| \geq 0.9784$ and $\|P\| \geq 1.9089$, which indicates that the above bounds are not sharp.

$$\Delta$$

The leading theorem in this section describes the upper and lower bounds for all the eigenvalues of the matrix P. Having obtained the estimates of all eigenvalues make it possible to estimate the determinant and trace of the matrix P. The results are similar to the lower bounds of all the eigenvalues of the continuous algebraic Lyapunov equation as obtained in (Karanam, 1981). The corresponding theorem is formulated and proved in (Mori et al., 1982a).

Theorem 3.1 *Let $\alpha_1, \alpha_2, \cdots, \alpha_n$ be the eigenvalues of P, $\beta_1, \beta_2, \cdots, \beta_n$ be the eigenvalues of Q, $\lambda_1, \lambda_2, \cdots, \lambda_n$ be the eigenvalues of A and $\sigma_1, \sigma_2, \cdots, \sigma_n$ be the eigenvalues of AA^T. It is assumed that the matrix A is asymptotically stable and $Q = Q^T > 0$ such that the solution $P = P^T > 0$ exists and is positive definite. Furthermore, it is assumed that*

$$\alpha_1 \geq \alpha_2 \geq \cdots \geq \alpha_n$$

$$\beta_1 \geq \beta_2 \geq \cdots \geq \beta_n$$

$$1 > \sigma_1 \geq \sigma_2 \geq \cdots \geq \sigma_n$$

and

$$Re[\lambda_1(A)] \geq Re[\lambda_2(A)] \geq \cdots \geq Re[\lambda_n(A)]$$

Then, the following inequalities hold for the eigenvalues of the matrix P

$$i) \quad \alpha_i \leq \beta_i + \frac{\sigma_1 \beta_1}{1 - \sigma_1}, \quad if \ \sigma_1 < 1$$

$$ii) \quad \alpha_i \geq \beta_i + \frac{\sigma_n \beta_n}{1 - \sigma_n}$$

for $i = 1, 2, \ldots, n$.

$$\square$$

The proof of this theorem is a direct consequence of the Ostrowski inequalities (Ostrowski, 1959) for the eigenvalues of a sum of symmetric matrices and for the eigenvalues of a matrix product.

Ostrowski Inequalities:

1) <u>Matrix Sum</u>: Let X and Y be symmetric matrices of dimension n. Then the eigenvalues satisfy

$$\lambda_i(X + Y) = \lambda_i(X) + h_i, \qquad i = 1, 2, ..., n$$

where

$$\lambda_1(Y) \geq h_i \geq \lambda_n(Y)$$

or by using notation from Theorem 3.1

$$\lambda_i(X) + \lambda_1(Y) \geq \lambda_i(X + Y) \geq \lambda_i(X) + \lambda_n(Y) \tag{3.12}$$

2) <u>Matrix Product</u>: Let $P = P^T$ and A be square matrices, then the eigenvalues satisfy

$$\lambda_i(A^T P A) = \lambda_i(A A^T P) = \theta_i \lambda_i(P), \quad \sigma_1^{1/2} \geq \theta_i \geq \sigma_n^{1/2} \tag{3.13}$$

Using these inequalities the reader can work out the proof of Theorem 3.1 quite easily.

Example 3.3: The solution matrix P of the algebraic discrete Lyapunov equation with A given in Example 3.1 and $Q = diag\{1, 2, 3, 4, 5\}$ has the following eigenvalues

$$\alpha_1 = 25.1666, \ \alpha_2 = 21.2278, \ \alpha_3 = 4.9507$$
$$\alpha_4 = 4.2603, \ \alpha_5 = 3.1623$$

It is obtained that $\sigma_1 = 1.4053$ and $\sigma_5 = 0.0390$. With $\beta_1 = 1$, $\beta_2 = 2$, $\beta_3 = 3$, $\beta_4 = 4$, $\beta_5 = 5$ the following lower bounds are obtained for the solution eigenvalues

$$\alpha_1 \geq 5.0406, \ \alpha_2 \geq 4.0406, \ \alpha_3 \geq 3.0406$$
$$\alpha_4 \geq 2.0406, \ \alpha_5 \geq 1.0406$$

Since $\sigma_1 > 1$ the upper bounds from Theorem 3.1 cannot be applied.

$$\Delta$$

In the paper of (Yasuda and Hirai, 1979) it has been shown that the extremal eigenvalues of the solution of the discrete algebraic Lyapunov equation satisfy

$$\alpha_1 \leq \frac{\beta_1}{1 - \sigma_1}, \qquad \alpha_n \geq \frac{\beta_n}{1 - \sigma_n}$$

which is the special case of the results stated in Theorem 3.1.

By using the Weyl matrix inequality, Mori has produced a generalization of results from Theorem 3.1, that is

$$i) \quad \alpha_i \leq \frac{\sigma_1^m \beta_1 + (1 - \sigma_1^m)\beta_{k+1}}{1 - \sigma_1}, \quad i + mk = n$$

$$\text{(3.14)}$$

$$ii) \quad \alpha_i \geq \frac{\sigma_n^m \beta_n + (1 - \sigma_n^m)\beta_{n-k}}{1 - \sigma_n}, \quad i - mk = n$$

where m and k are nonnegative integers. It has been shown (Mori, 1983), on a simple example, that for some choices of m and k the bounds (3.14) are sharper than the corresponding ones from Theorem 3.1. However, in general, there is no way to find the best possible choices for m and k, due to problem nonlinearities, so that one is left only with some heuristic search methods (Mori, 1983).

The result stated in (3.14) can be established by using the Weyl inequality, which can be found in (Amir-Moez, 1956).

Weyl Inequality:

Let X and Y be hermitian matrices of dimension n. Then, the eigenvalues of a matrix sum satisfy

$$\lambda_{i+j-n}(X + Y) \geq \lambda_i(X) + \lambda_j(Y), \quad i, j = 1, 2, ..., n$$
$$i + j \geq n + 1$$

$$\text{(3.15)}$$

$$\lambda_{i+j-1}(X + Y) \leq \lambda_i(X) + \lambda_j(Y), \quad i, j = 1, 2, ..., n$$
$$i + j \leq n + 1$$

It is well known that the discrete-time algebraic Lyapunov equation has a unique positive definite solution even in the case when the matrix

Q is positive semi-definite. This is possible under the assumption that the controllability matrix defined by

$$M = [B \quad AB \quad \dots \quad A^{n-1}B], \quad Q = BB^T \tag{3.16}$$

has full rank (Chen, 1984). Note that the controllability matrix in (3.16) corresponds to the controllability matrix of the transpose of (3.1), that is, to

$$APA^T + Q = P \tag{3.17}$$

The next theorem is an extension of Theorem 2.7 (Mori et al., 1986) for the continuous algebraic Lyapunov equation. This theorem produces the estimates of the eigenvalue bounds in terms of the controllability matrix (Mori et al., 1985).

Theorem 3.2 *The solution of (3.17) satisfies*

$$\lambda_{min}(MM^T)P_1 \le P \le \lambda_{max}(MM^T)P_1$$

$$\lambda_{min}(MM^T)\lambda_i(P_1) \le \lambda_i(P) \le \lambda_{max}(MM^T)\lambda_i(P_1)$$

where P_1 is the solution of the algebraic Lyapunov matrix equation

$$A^n P_1 (A^T)^n - P_1 + I = 0$$

and I is identity. Furthermore, if A is normal, that is, $A^T A = AA^T$, we have

$$\lambda_{min}(MM^T)\lambda_i[(I - AA^T)^n]^{-1} \le \lambda_i(P) \le$$
$$\lambda_{max}(MM^T)\lambda_i[(I - AA^T)^n]^{-1}$$

$\square$

The nontrivial bounds $(\lambda_{min}(MM^T) > 0)$ given in Theorem 3.2 can be improved in the case when A has multiple eigenvalues (Troch, 1987).

Theorem 3.3 *Let the number of distinct eigenvalues of A be $m \le n$, and $\hat{M}$ be given by*

$$\hat{M}^T = [C^T \quad A^T C^T \quad \dots \quad (A^T)^{m-1}C^T]$$

Define the matrix $\hat{G}$ by

$$\hat{G} = (g_{ij}), \quad g_{ij} = \sum_{k=0}^{\infty} a_i(k)a_j(k)$$

where $a_i(k), i = 1, 2, \ldots, m$ are obtained from

$$A^k = \sum_{i=0}^{m-1} a_i(k)A^i$$

then

$$\lambda_{min}(\hat{G})\hat{M}\hat{M}^T \leq P \leq \lambda_{max}(\hat{G})\hat{M}\hat{M}^T$$

$$\lambda_{min}(\hat{G})\lambda_i(\hat{M}\hat{M}^T) \leq \lambda_i(P) \leq \lambda_{max}(\hat{G})\lambda_i(\hat{M}\hat{M}^T), \ i = 1, 2, \ldots, n$$

$\square$

On the contrary to Theorem 3.2, here one needs to use only the minimal polynomial instead of the characteristic polynomial in order to get $a_i(k)'s$ so that the matrix $\hat{G}$ is simpler in this case, which reduces the required calculations. In addition, the obtained bounds are sharper than those from Theorem 3.2 (Troch, 1987). In a particular example done in (Troch, 1987) the lower and upper bounds coincide.

Eigenvalue summation and product bounds of the solution matrix of the discrete-time algebraic Lyapunov equation are given at the end of this section since they can be also used to determine the bounds for the solution's trace and determinant.

3.2.2 Trace Bounds

The first result on the trace bounds of the solution matrix of the discrete algebraic Lyapunov equation was reported in (Mori et al., 1982b), where the term arithmetic mean is used for the following quantity

$$m_a(X) = \frac{1}{n} tr(X) \tag{3.18}$$

The results for both lower and upper bounds are described in the following theorem.

Theorem 3.4 *The arithmetic mean of the positive definite matrix P in equation (3.1) satisfies the following inequalities*

$$m_a(P) \geq \frac{m_a(Q)}{1 - \lambda_{min}(AA^T)}$$

$$m_a(P) \leq \frac{m_a(Q)}{1 - \lambda_{max}(AA^T)} \qquad if \ \|A\| < 1$$

where $\|A\| = \lambda_{max}^{1/2}(AA^T)$.

□

The proof of this theorem follows directly by using the following known inequality (Kleinman and Athans, 1968).

Trace of the Product Inequality:

For any two positive semi-definite matrices $X = X^T \geq 0$ and $Y = Y^T \geq 0$ we have

$$\lambda_{min}(X)tr(Y) \leq tr(XY) \leq \lambda_{max}(X)tr(Y) \qquad (3.19)$$

Note that it was shown in (Sanjuk and Rhodes, 1987) that the above inequality is valid even for $Y = Y^T \geq 0$ and X any square matrix. A stronger result of this important matrix inequality than the one in (3.19) is also obtained in (Mori, 1988; Hmamed, 1989) — see Appendix.

Example 3.4: For the matrices A and Q as in Example 3.1, Theorem 3.4 gives the following lower bound

$$m_a(P) \geq 1.0008$$

while the exact value is $m_a(P) = 4.4443$. The upper bound is not applicable since $\|A\| = 1.4053$ is greater than one.

Δ

The next theorem presents a lower bound of $tr(P)$ which is tighter than the above bound when $Q = kI$ ($k > 0$), (Kwon et al., 1985).

Theorem 3.5 *The trace of the positive definite matrix P satisfying (3.1) has the following lower bound*

$$tr(P) \geq \frac{n^2 \lambda_{min}(Q)}{n - \sum_{i=1}^{n} |\lambda_i(A)|^2}$$

□

This result is an extension of the third inequality given in Theorem 2.11. In order to demonstrate that the bound in Theorem 3.5 is tighter than that in Theorem 3.4, we notice that for the case when $Q = kI$ the bound in Theorem 3.5 becomes

$$tr(P) \geq \frac{n^2 k}{n - \sum_{i=1}^{n} |\lambda_i(A)|^2}$$

and the bound in Theorem 3.4 is

$$tr(P) \geq \frac{nk}{1 - \lambda_{min}(AA^T)}$$

Due to the fact that $|\lambda_i(A)|^2 \geq \lambda_{min}(AA^T), i = 1, 2, \ldots, n$, it is true that the bound in Theorem 3.5 is tighter than that in Theorem 3.4.

Example 3.5: (Kwon et al., 1985). Let

$$A = \begin{bmatrix} 0 & 0.1 \\ -0.2 & -0.3 \end{bmatrix}, \quad Q = \begin{bmatrix} 1 & 0 \\ 0 & 1 \end{bmatrix}$$

Then solving the discrete algebraic Lyapunov equation (3.1), we get

$$P = \begin{bmatrix} 1.044 & 0.065 \\ 0.065 & 1.106 \end{bmatrix}$$

so that $tr(P) = 2.150$. On the other hand, from Theorem 3.5 we have

$$tr(P) \geq 2.051$$

whereas from Theorem 3.4 we get

$$tr(P) \geq 2.006$$

$$\Delta$$

In the recent paper (Komaroff, 1992) the following lower trace bound is obtained

$$tr(P) \geq n \left[|Q| \prod_{i=1}^{n} \left(1 - |\lambda_i(A)|^2 \right)^{-1} \right]^{1/n} \tag{3.20}$$

It is shown in (Komaroff, 1992) that this bound is in general tighter than the one from Theorem 3.5. It is also shown that (3.20) is tighter than the corresponding bound from Theorem 3.4 in the special case when $Q = I$.

Another lower trace bound, which is better than one from (Kwon et al., 1985), has been derived in (Komaroff and Shahian, 1992)

$$tr(P) \geq \frac{\left[tr\left(Q^{1/2}\right)\right]^2}{n - \sum\limits_{i=1}^{n} |\lambda_i(A)|^2} \tag{3.21}$$

However, no general statement can be made whether or not this bound is sharper than the one obtained by (Mori et al., 1982b; Komaroff 1992).

Example 3.6: Let us compare results from Theorem 3.5 and formulas (3.20)-(3.21) on a real physical system. A discrete-time model of an L-1011 fighter aircraft is obtained from its continuous-time model (Beale and Shafai, 1989) by discretizing it using MATLAB function c2d with the sampling period $T = 0.5$. The corresponding discrete-time system matrix is obtained as

$$A = \begin{bmatrix} 0.9971 & 0.3228 & 0.0825 & -0.4662 \\ -0.0158 & 0.3846 & 0.3414 & -1.4869 \\ 0.0062 & -0.0037 & 0.1159 & 0.5449 \\ 0.0150 & 0.0044 & -0.2178 & 0.7247 \end{bmatrix}$$

This matrix is asymptotically stable with the eigenvalues given by 0.9507, $0.4533 \pm j0.1476$, 0.3650. Taking for the Q matrix

$$Q = \begin{bmatrix} 1 & 0 & 1 & 0 \\ 0 & 2 & 0 & 1 \\ 1 & 0 & 3 & 0 \\ 0 & 1 & 0 & 1 \end{bmatrix}$$

whose eigenvalues are 3.4142, 2.6180, 0.5858, 0.3820, we get the solution matrix P from (3.1) as

$$P = \begin{bmatrix} 13.6776 & 6.4429 & 11.5523 & -29.5081 \\ 6.4429 & 5.6864 & 5.9740 & -16.1979 \\ 11.5523 & 5.9740 & 14.4755 & -31.8487 \\ -29.5081 & -29.5081 & -31.8487 & 92.4851 \end{bmatrix}$$

The trace of P is given by $tr(P) = 126.3247$. However, the results from Theorem 3.5, (3.20), and (3.21) give the lower bounds for the trace of P far away from the actual value, that is

$$Theorem\ 3.5 \Rightarrow tr(P) \geq 2.4366$$
$$formula\ (3.20) \Rightarrow tr(P) \geq 10.0713$$
$$formula\ (3.21) \Rightarrow tr(P) \geq 9.3744$$

This example indicates that the additional research is needed to improve the lower trace bounds of the solution matrix P of the discrete-time algebraic Lyapunov equation.

$$\Delta$$

An upper bound for the trace of P will be given at the end of this section, where an eigenvalue summation upper bound is presented.

3.2.3 Determinant Bounds

A result for determinant bounds, which has the similar form to trace bounds given in Theorem 3.4, is presented in (Mori et al., 1982b), where the geometric mean is used for the following quantity

$$m_g(X) = |X|^{1/n}$$

The result establishes both lower and upper bounds for the determinant of the geometric mean in the next theorem.

Theorem 3.6 *The geometric mean of the positive definite matrix P in equation (3.1) satisfies the following inequalities*

$$(i)\quad m_g(P) \geq \frac{m_g(Q)}{1 - m_g^2(A)}$$

and if $\|A\| = \lambda_{max}^{1/2}(A^T A) < 1$ and Q is chosen such that

$$\|A\|^2 Q \geq A^T Q A$$

then, the upper bound is given by

$$(ii)\quad m_g(P) \leq \frac{m_g(Q)}{1 - \lambda_{max}(AA^T)}$$

$$\square$$

Note that the inequality (i) is a direct consequence of the very well-known Minkovski inequality for determinants (Marcus and Minc, 1964), which is given below.

Minkovski Inequality:
For any two square matrices the following property of determinants holds

$$|X + Y|^{1/n} \geq |X|^{1/n} + |Y|^{1/n} \tag{3.22}$$

Example 3.7: For the steam power system from Example 3.1 the lower bound is given by

$$m_g(P) \geq 1.08014$$

while the exact value is $m_g(P) = 2.5255$.

$\triangle$

By using a numerical example (Tran and Sawan, 1984) have demonstrated that the better result can be obtained for the lower determinant bound than the one in Theorem 3.6. Their result is given in the following theorem.

Theorem 3.7 *The determinant of the positive definite solution matrix P of equation (3.1) satisfies the inequality*

$$|P| \geq |Q| \left[\frac{n}{n - \sum\limits_{i=1}^{n} |\lambda_i(A)|^2} \right]^n$$

$\square$

Proof: The proof of this theorem is based on the inequalities from (Bechenbach and Bellman, 1965; Patel and Toda 1979).

Inequality A: Let $H \in \Re^{n \times n}$ and $L \in \Re^{n \times n}$ with $L > 0$. Then the trace operator satisfies

$$tr\left(L^{-1} H L H^T\right) \geq \sum_{i=1}^{n} |\lambda_i(H)|^2 \geq [tr(H)]^2 \tag{3.23}$$

Inequality B: Let $R = R^T \in \Re^{n \times n}$ and $S = S^T \in \Re^{n \times n}$ with $R > 0, S > 0$. Then, the determinant of R satisfies

$$|R|^{1/n} = \min_{|S|=1} \left\{ \frac{tr(RS)}{n} \right\} \tag{3.24}$$

In order to complete the proof, we multiply (3.1) from the left by P^{-1} and take the trace of the obtained equation which leads to

$$tr(I) = n = tr\left(P^{-1}A^T PA + P^{-1}Q\right)$$

Applying Inequality A to this equation with $L = P$ and $H = A^T$, we get

$$tr\left(P^{-1}Q\right) \leq n - \sum_{i=1}^{n} |\lambda_i(A)|^2$$

Using now Inequality B with $R = P^{-1}, S = \frac{Q}{|Q|^{1/n}}$, we get

$$n\left|Q^{1/n}\right|\left|P^{-1}\right|^{1/n} \leq n - \sum_{i=1}^{n} |\lambda_i(A)|^2$$

which easily leads to the stated result of Theorem 3.7.

∎

This theorem does not assume that A is stable matrix, but in such a case if the denominator is zero, we cannot use this bound. For asymptotically stable A this problem is removed. Assuming the stability of the matrix A in the follow-up paper (Fu and Sawan, 1985) have analytically established that the bound from Theorem 3.7 is sharper than the corresponding bound from Theorem 3.6. The analytical proof of this fact has been also given in (Garloff, 1986).

In the papers by (Hmamed, 1991; Komaroff, 1992), a tighter result is obtained for the determinant lower bound than in (Tran and Sawan, 1984). The result is given below in the following theorem.

Theorem 3.8 *The lower bound of the determinant of the solution matrix P of the discrete-time algebraic Lyapunov equation is given by*

$$|P| \geq |Q| \prod_{i=1}^{n} \left(1 - |\lambda_i(A)|^2\right)^{-1}$$

□

Example 3.8: A discretized model of hydroturbine governors (Arnautovic and Skataric, 1991) is obtained using MATLAB and its function c2d with $T = 1$ as

$$A = \begin{bmatrix} 0.4916 & 0 & 0 & 0 & 0 \\ 0 & 0.1353 & 0 & 0 & 0 \\ 0.2104 & 0.2283 & 0.2343 & 0.0319 & -0.0013 \\ -0.0086 & -0.0148 & 0.0316 & -0.4563 & -0.0164 \\ -0.3176 & -0.6624 & 1.8003 & 22.4148 & -0.4147 \end{bmatrix}$$

For $Q = I_5$, it has been found using MATLAB functions dlyap and det that $det(P) = 1167.022$. However, Theorems 3.7 and 3.8 have produced much lower values, respectively given by $det(P) \geq 5.3478$ and $det(P) \geq 7.1853$. This indicates necessity for further research on the lower bound of the determinant of the matrix P.

It is interesting to observe that the matrix A has the singularly perturbed form (see Chapter 5), which suggests that the singular perturbation methodology might be used in an attempt to achieve sharper bounds for the main attributes of the solution matrix P for linear dynamical systems having separation of eigenvalues into two disjoint groups causing slow and fast time phenomena.

$\triangle$

Summation and Product Eigenvalue Bounds

Finally, at the end of this section, we want to point out that a generalization of some of the previous bounds has been obtained by (Garloff, 1986) in the form of the following theorem, which in fact, gives simultaneous summation and product eigenvalue bounds.

Theorem 3.9 *The eigenvalues of the solution of the discrete-time algebraic Lyapunov equation satisfy*

$$i) \quad \lambda_i(P) \geq \lambda_i(Q), \qquad i = 1, 2, ..., n$$

$$ii) \quad \lambda_1(P) \geq \sum_{i=1}^{k} \lambda_{n-i+1}(Q) \left[k - \sum_{i=1}^{k} |\lambda_i(A)|^2 \right], \quad k \leq n$$

$$iii) \quad \prod_{i=1}^{k} \lambda_i(P) \geq \prod_{i=1}^{k} \lambda_{n-i+1}(Q) \left[k \left(k - \sum_{i=1}^{k} |\lambda_i(A)|^2 \right)^{-1} \right]^k, \quad k \leq n$$

$$iv) \quad \sum_{i=1}^{k} \lambda_i(P) \leq \sum_{i=1}^{k} \lambda_i(Q) \left[1 - \sigma_1^2(A)\right]^{-1}, \quad k \leq n, \;\; if \;\; \sigma_1(A) < 1$$

$$v) \quad \sum_{i=1}^{k} \lambda_{n-i+1}(P) \geq \sum_{i=1}^{k} \lambda_{n-i+1}(Q) \left[1 - \sigma_n^2(A)\right]^{-1}, \quad k \leq n$$

$\square$

These inequalities are derived in the context of a general study of bounds for the discrete-time algebraic Riccati equation so that the reader is referred to the original work (Garloff, 1986) for the corresponding proofs. The work of (Garloff, 1986) on the simultaneous eigenvalue lower bounds is complemented in the paper by (Hmamed, 1991).

It is important to point out that from inequality iv) from Theorem 3.9, we have the upper bound for the trace of the matrix P as follows:

$$tr(P) \leq \frac{tr(Q)}{1 - \sigma_1^2(A)} \quad if \;\; \sigma_1(A) = \lambda_1^{1/2}(AA^T) < 1 \qquad (3.25)$$

Also, inequality iii) for $k = n$ implies the result stated in Theorem 3.7.

Similarly to inequality v) from the previous theorem, the lower summation bound is obtained in (Komaroff and Shahian, 1992) as

$$\sum_{i=1}^{k} \lambda_{n-i+1}(P) \geq k^2 \lambda_n(Q) \sum_{i=1}^{k} \left(1 - |\lambda_i(A)|^2\right)^{-1}, \quad k \leq n \qquad (3.26)$$

Some additional summation bounds can be found in (Hmamed, 1990), where the solution matrix bounds have been studied for the differential and difference Lyapunov equations. The obtained results in the limit (steady state) also give summation bounds for the solution's attributes of the continuous and discrete algebraic Lyapunov equations. These results will be presented in Chapter 4, where we study the differential and difference Lyapunov equations.

Note that it is almost impossible to compare different bounds since they have been obtained using different approaches and different inequalities, which is indicated by T. Mori, a leading researcher in the theory of bounds of the Lyapunov equations: "With few exceptions, a general comparison between any parallel bounds for the same measure is either not easy or actually impossible" (Mori and Derese, 1984). Similar conclusion follows from the papers by Komaroff.

The unified continuous-discrete-time approach to linear system theory is introduced in (Middleton and Goodwin, 1990), where the so-called unified continuous-discrete-time algebraic Lyapunov equation is presented. The eigenvalue bounds of the unified algebraic Lyapunov equation are studied in (Mrabti and Hmamed, 1992).

3.3 Numerical Solutions

Like in the case of the algebraic Lyapunov equation in continuous-time, one may solve the discrete-time algebraic Lyapunov equation by using a direct approach, that is, by solving a system of linear algebraic equations with $0.5n(n+1)$ unknowns. These equations can obtained, for example, by using the bilinear transformation

$$B = (A - I)^{-1}(A + I)$$

to transform (3.1) into

$$B^T P_b + P_b B + Q = 0$$

and then by solving the last equation by any of the direct methods described in Section 2.1.1. But this yields $O(n^6)$ multiplications, and therefore, it is impractical for large systems. The other approach is to use iterative methods similar to the ones used for continuous case, which can reduce the number of multiplications to $O(n^3)$, (Pace and Barnett, 1972). An iterative algorithm for solving discrete algebraic Lyapunov equation dual to Algorithm 2.4 (Davison and Man, 1968) is presented in Chapter 7 as Algorithm 7.1, (Smith, 1968). In fact, most of the numerical methods to solve (3.1), direct or iterative, use such a kind of transformation and then utilize one of the algorithms from Section 2.3 to solve the transformed equation. Since these procedures only have the additional steps of forward and backward transformation as compared to the solutions for continuous case, therefore, in this section only those methods which do not involve such a transformation will be discussed.

The method to be presented is described in (Barraud, 1977) and is similar to the Bartels-Stewart algorithm. Let W be an orthogonal matrix such that

$$\widetilde{A} = W^T A W$$

be in real upper Schur form. The matrix W can be obtained by first transforming A to upper Hessenberg form using Householder's method, and then by using the QR algorithm (Wilkinson, 1965) to get Schur's form (Martin et al., 1970), like in the development of Algorithm 2.6. Defining the following matrices

$$\widetilde{P} = W^T P W$$

and

$$\widetilde{Q} = -W^T Q W$$

It is easy to see that (3.1) has the equivalent form

$$\widetilde{A}^T \widetilde{P} \widetilde{A} - \widetilde{P} = \widetilde{Q} \qquad (3.27)$$

The structures of $\widetilde{A}, \widetilde{Q}$, and $\widetilde{P}$ are similar to (2.54)-(2.56), respectively. Due to the fact that the partitions of $\widetilde{A}, \widetilde{Q}$, and $\widetilde{P}$ are conformal, we can construct the following equations. For $l = 1, 2, \ldots, p$

$$\sum_{i=1}^{k} \left(\widetilde{A}_{ik} \right)^T \left(\sum_{j=1}^{l} \widetilde{P}_{ij} \widetilde{A}_{jl} \right) - \widetilde{P}_{kl} = \widetilde{Q}_{kl}; \quad k = l, l+1, \ldots, p \qquad (3.28)$$

These equations may be solved successively for $\widetilde{P}_{11}, \widetilde{P}_{21}, \ldots, \widetilde{P}_{p1}, \widetilde{P}_{22}, \ldots,$ $\widetilde{P}_{22}, \ldots, \widetilde{P}_{pp}$. Each such block $\widetilde{P}_{kl}$ involves the solution of a matrix equation of the form (3.1), but since all the submatrices $\widetilde{A}_{kl}$ are of order at most two by construction, then one has to solve a system of linear equations of order at most four. If $\widetilde{A}_{kl}$ is of order one, then (3.28) represents a scalar equation whose solution is trivial. If $k = l$, then (3.28) represents only three equations in three unknown values of symmetric matrices $\widetilde{P}_{kk}, k = 1, 2, \ldots, p$. Several other cases are given in (Barraud,

1977), which can be utilized to cut down the number of multiplications for the solution of (3.28). Finally, the original solution P is given by

$$P = W\widetilde{P}W^T$$

To summarize, the algorithm has following steps.

Algorithm 3.1:

1. Transform A to real upper Schur form $\widetilde{A} = W^T AW$.
2. Construct $\widetilde{Q} = -W^T QW$, and $\widetilde{P} = W^T PW$.
3. Partition $\widetilde{A}, \widetilde{Q}$, and $\widetilde{P}$ according to (2.54)-(2.56), respectively.
4. For $l = 1, 2, \ldots, p$ and $k = l, l+1, \ldots, p$ solve (3.28) to form the matrix $\widetilde{P}$.
5. Construct the required solution by $P = W\widetilde{P}W^T$.

$$\Delta$$

Note that Algorithm 3.1 does not require the asymptotic stability of the matrix A.

This algorithm requires $2.5n^2 + n/2$ storage locations and $4n^3(\sigma+1)$ multiplications as in the Bartels-Stewart algorithm.

A numerical method for solving (3.1) based on the QR algorithm, which is similar to Algorithm 3.1, is developed in (Kitagawa, 1977). The Hessenberg/Schur technique for numerical solution of the discrete algebraic Lyapunov equation is also discussed in (Golub et al., 1979; Hammarling, 1982, 1991; Varga, 1990).

In the case when the system matrix A is in companion form a numerical algorithm for solving the corresponding discrete algebraic Lyapunov equation is given in (Berger, 1971). The matrix sign method for discrete algebraic Lyapunov equation is considered in (Denman and Beavers, 1976; Roberts, 1980).

3.4 Summary

The procedures for solving the discrete algebraic Lyapunov equation are similar to those for continuous algebraic Lyapunov equation. Two different bilinear transformations are presented, which convert the discrete algebraic Lyapunov equation into the continuous algebraic Lyapunov equation. None of them is preferred on the other. Once the transformed continuous algebraic Lyapunov equation is solved, by some of the methods described in Chapter 2, it can be transformed back to have the solution for the discrete algebraic Lyapunov equation. The skew-symmetric matrix method is used in order to reduce the number of linear equations from $0.5n(n + 1)$ to $0.5n(n - 1)$. A transformation approach is also discussed, which solves the discrete algebraic Lyapunov equation without converting it into the continuous Lyapunov equation. This approach transforms the coefficient matrix to its Jordan canonical form and then the final solution is calculated in terms of the Jordan blocks.

The bounds for different parameters of the solution matrix are described including eigenvalue bounds, trace bounds, and determinant bounds. These bounds can be used to examine behavior of the underlying system without solving the discrete Lyapunov equation. Some examples are also given in order to show the tightness and efficiency of the bounds. It is important to point out that the final results on the best possible bounds for the solution matrix of the discrete algebraic Lyapunov equation are not obtained yet, so that further research is needed in that direction.

In the section on numerical solutions only one algorithm is described which is similar to Algorithm 2.6 from Chapter 2. Because of the wide use of Algorithm 2.6, similar kind of procedure and similar linear equations for discrete Lyapunov equation are presented. These equations can be solved sequentially to obtain the required solution matrix.

3.5 References

1. Amir-Moez, A., "Extreme properties of a Hermitian transformation and singular values of the sum and product of linear transformation," *Duke Math., J.*, vol.23, 463–476, 1956.

2. Arnautovic, D. and D. Skataric, "Suboptimal design of hydroturbine governors," *IEEE Trans. Energy Conversion*, vol.6, 438–444, 1991.

3. Barnett, S., "Simplification of Lyapunov matrix equation $A^T P A - P = -Q$," *IEEE Trans. Automatic Control*, vol.19, 446-447, 1974.

4. Barnett, S. and C. Storey, "The Liapunov matrix equation and Schwarz's form," *IEEE Trans. Automatic Control*, vol.12, 117–118, 1967.

5. Barnett, S. and C. Storey, *Matrix Methods in Stability Theory*, Nelson, London, 1970.

6. Barraud, A., "A numerical algorithm to solve $A^T X A - X = Q$," *IEEE Trans. Automatic Control*, vol.22, 883–885, 1977.

7. Beale, S. and B. Shafai, "Robust control system design with a proportional-integral controller," *Int. J. Control*, vol.50, 97–111, 1989.

8. Bechenbach, E. and R. Bellman, *Inequalities*, Springer Verlag, Berlin, 1965.

9. Berger, C., "A numerical solution of the matrix equation $P = \Phi P \Phi^T + S$," *IEEE Trans. Automatic Control*, vol.16, 381-382, 1971.

10. Bitmead, R. and H. Weiss, "On the solution of the discrete-time Lyapunov matrix equation in controllable canonical form," *IEEE Trans. Automatic Control*, vol.24, 481–482, 1979.

11. Chen, C., *Linear System Theory and Design*, Holt, Rinehart and Winston, New York, 1984.

12. Davison, E. and F. Man, "The numerical solution of $A^T Q + QA = -C$," *IEEE Trans. Automatic Control*, vol.13, 448-449, 1968.

13. Denman, E. and A. Beavers, "The matrix sign function and computations in systems," *Appl. Math. and Computation*, vol.2, 63–94, 1976.

14. Fan, K., "On a theorem of Weyl concerning eigenvalues of linear transformations," *Proc. National Academy of Science, USA,* vol.35, 652–655, 1949.

15. Fu, S. and M. Sawan, "Solution of the discrete Lyapunov equation," *SIAM J. Alg. Disc. Meth.*, vol.6, 341–344, 1985.

16. Garloff, J., "Bounds for eigenvalues of the solution of the discrete Riccati and Lyapunov equations and the continuous Lyapunov equation," *Int. J. Control*, vol.43, 423-431, 1986.

17. Golub, G., S. Nash, and C. Loan, "A Hessenberg-Schur method for the problem $AX + XB = C$," *IEEE Trans. Automatic Control*, vol.24, 909-913, 1979.

18. Gomathi, K., S. Prabhu, and M. Pai, "A suboptimal controller for minimum sensitivity of closed-loop eigenvalues to parameter variations," *IEEE Trans. Automatic Control*, vol.25, 587–588, 1980.

19. Hammarling, S., "Numerical solution of the stable, non-negative definite Lyapunov equation," *IMA J. Numerical Analysis*, vol.2, 303–323, 1982.

20. Hammarling, S., "Numerical solution of the discrete-time convergent, non-negative definite Lyapunov equation," *Systems & Control Letters*, vol.17, 137–139, 1991.

21. Heinen, J., "A technique for solving the extended discrete Lyapunov matrix equation," *IEEE Trans. Automatic Control*, vol.17, 156-157, 1972.

22. Hmamed, A., "A matrix inequality," *Int. J. Control*, vol.49, 363-365, 1989.

23. Hmamed, A., "Differential and difference Lyapunov equations: simultaneous eigenvalue bounds," *Int. J. Systems Science*, vol.21, 1335-1344, 1990.

24. Hmamed, A., "Discrete Lyapunov equation: simultaneous eigenvalue lower bounds," *Int. J. Systems Science*, vol.22, 1121-1126, 1991.

25. Kailath, T., *Linear Systems*, Prentice Hall, Englewood Cliffs, 1980.

26. Karanam, V., "Lower bounds on the solution of Lyapunov matrix and algebraic Riccati equation," *IEEE Trans. Automatic Control*, vol.26, 1288-1290, 1981.

27. Kitagawa, G., "An algorithm for solving the matrix equation $X = FXF^T + S$," *Int. J. Control*, vol.25, 745–753, 1977.

28. Kleinman, D. and M. Athans, "The design of suboptimal time-varying systems," *IEEE Trans. Automatic Control*, vol.13, 150–159, 1968.

29. Komaroff, N., "Lower bounds on the solution of the discrete algebraic Lyapunov equations," *IEEE Trans. Automatic Control*, vol.37, 1017–1018, 1992.

30. Komaroff, N. and B. Shahian, "Lower summation bounds for the discrete Riccati and Lyapunov equations," *IEEE Trans. Automatic Control*, vol.37, 1078–1080, 1992.

31. Kwon, W. and A. Pearson, "A note on algebraic matrix Riccati equation," *IEEE Trans. Automatic Control*, vol.22, 143-144, 1977.

32. Kwon, B., M. Youn, and Z. Bien, "On bounds of the Riccati and Lyapunov matrix equations," *IEEE Trans. Automatic Control*, vol.30, 1134-1135, 1985.

33. Mahmoud, M., Y. Chen, and M. Singh, "Discrete two-time-scale systems," *Int. J. Systems Science*, vol.17, 1187-1207, 1986.

34. Marcus, M. and H. Minc, *A Survey of Matrix Theory and Inequalities*, Allyn and Bacon, Boston, 1964.

35. Martin, R., G. Peters, and J. Wilkinson, "The QR algorithm for real Hessenberg matrices," *Numer. Math.*, vol.14, 219–231, 1970.

36. Middleton, R. and G. Goodwin, *Digital Control and Estimation: A Unified Approach*, Prentice Hall, Englewood Cliffs, 1990.

37. Mori, T. "Note on the existence regions for eigenvalues of the solution to the discrete Lyapunov matrix equation," *Int. J. Systems Science*, vol.14, 463–466, 1983.

38. Mori, T., "Comments on 'A matrix inequality associated with bounds on solutions of algebraic Riccati and Lyapunov equations'," *IEEE Trans. Automatic Control*, vol.33, 1088, 1988.

39. Mori, T., N. Fukuma, and M. Kuwahara, "Upper and lower bounds for the solution to the discrete Lyapunov matrix equation," *Int. J. Control*, vol.36, 889-892, 1982a.

40. Mori, T., N. Fukuma, and M. Kuwahara, "On the discrete Lyapunov matrix equation," *IEEE Trans. Automatic Control*, vol.27, 463-464, 1982b.

41. Mori, T. and I. Derese, "A brief summary of the bounds on the solution of the algebraic matrix equations in control theory," *Int. J. Control*, vol.39, 247-256, 1984.

42. Mori, T., N. Fukuma, and M. Kuwahara, "Eigenvalue bounds for the discrete Lyapunov matrix equation," *IEEE Trans. Automatic Control*, vol.30, 925-926, 1985.

43. Mori, T., N. Fukuma, and M. Kuwahara, "Explicit solution and eigenvalue bounds in the Lyapunov matrix equation," *IEEE Trans. Automatic Control*, vol.31, 656-658, 1986.

44. Mrabti, M and A. Hmamed, "Bounds for the solution of the Lyapunov matrix equation — A unified approach," *Systems & Control Letters*, vol.18, 73–81, 1992.

45. Ostrowski, A., "A quantitative formulation of Sylvester's law of inertia," *Proc. National Academy of Science, USA*, vol.45, 740–744, 1959.

46. Pace, I. and S. Barnett, "Comparison of numerical methods for solving Liapunov matrix equations," *Int. J. Control*, vol.15, 907-915, 1972.

47. Patel, R. and M. Toda, "On norm bounds for algebraic Riccati and Lyapunov equations," *IEEE Trans. Automatic Control*, vol.23, 87-88, 1978.

48. Patel, R. and M. Toda, "Trace inequalities involving hermitian matrices," *Linear Algebra and Its Appl.*, vol.23, 13–20, 1979.

49. Popov, V., "Hyperstability and optimality of automatic systems with several control functions," *Rev. Roum. Sci. Tech.*, vol.9, 629-690, 1964.

50. Power, H., "Equivalence of Lyapunov matrix equations for continuous and discrete systems," *Electronics Letters*, vol.3, 83, 1967.

51. Power, H., "Two applications of the matrix transformation $A = (B+I)(B-I)^{-1}$," *Electronics Letters*, vol.4, 479–481, 1968.

52. Power, H., "A note on matrix equation $A^T LA - L = -K$," *IEEE Trans. Automatic Control*, vol.14, 411-412, 1969.

53. Roberts, J., "Linear model reduction and solution of the algebraic Riccati equation by use of the sign function," *Int. J. Control*, vol.32, 677–687, 1980.

54. Sarma, I. and M. Pai, "A note on the Lyapunov matrix equation for linear discrete systems," *IEEE Trans. Automatic Control*, vol.13, 119–121, 1968.

55. Sanjuk, J. and I. Rhodes, "A matrix inequality associated with bounds on solutions of algebraic Riccati and Lyapunov equations," *IEEE Trans. Automatic Control*, vol.32, 739, 1987.

56. Smith, R., "Matrix equation $XA + BX = C$," *SIAM J. Appl. Math.*, vol.16, 198-201, 1968.

57. Tran, M. and M. Sawan, "Correction to 'A note on discrete Lyapunov and Riccati matrix equations'," *Int. J. Control*, vol.46, 1119, 1987.

58. Troch, I., "Improved bounds for eigenvalues of solution of Lyapunov equations," *IEEE Trans. Automatic Control*, vol.32, 744-747, 1987.

59. Troch, I., "Solving the discrete Lyapunov equation using the solution of the corresponding continuous Lyapunov equation and vice versa," *IEEE Trans. Automatic Control*, vol.33, 944–946, 1988.

60. Varga, A., "A note on Hammarling's algorithm for the discrete Lyapunov equation," *Systems & Control Letters*, vol.15, 273–275, 1990.

61. Weyl, H., "Inequalities between the two kinds of eigenvalues of linear transformation," *Proc. National Academy of Science, USA*, vol.35, 408–411, 1949.

62. Wilkinson, J., *The Algebraic Eigenvalue Problem*, Oxford University Press, Cambridge, 1965.

63. Yasuda, K. and K. Hirai, "Upper and lower bounds on solution of algebraic Riccati equation," *IEEE Trans. Automatic Control*, vol.24, 483-487, 1979.

Chapter Four

Differential and Difference
Lyapunov Equations

In this chapter, we present the explicit solutions, bounds of the solution main attributes (eigenvalues, trace, determinant), and numerical solutions of the differential and difference Lyapunov equations.

The explicit solutions have been presented for diagonalizable matrix A only. The bounds considered are based on a few results available in the literature (Mori et al., 1986, 1987; Hmamed, 1990). We also present the solution bounds for the differential Lyapunov equation according to the work of (Geromel and Bernussou, 1979). The obtained solution bounds at steady state give the bounds for the solution of the continuous algebraic Lyapunov equation.

Numerical solutions present mostly work of (Subrahmanyam, 1986) for the differential time varying Lyapunov equation, and the classic work of (Davison, 1975), (heavily based on Algorithm 2.4 (Davison and Man, 1968)), which is applicable to the corresponding time invariant equation.

In addition, we consider the differential and difference Lyapunov equations with small parameters corresponding to singularly perturbed and weakly coupled systems (Gajic et al., 1990; Gajic and Shen, 1993), where the order-reduction is achieved by using suitable state transformations so that the original (global) Lyapunov equations are decomposed into the reduced-order subsystem local Lyapunov equations.

At the end of this chapter, the coupled differential Lyapunov equations are presented according to the work of (Jodar and Mariton, 1987).

4.1 Explicit Solutions

In this section we introduce the differential and difference Lyapunov equations, state main existence results, and present a method for the explicit solution of a class of time invariant differential Lyapunov equations. Since the solution of the difference Lyapunov equation can be obtained by a simple recursion, the corresponding recursive formula represents the required explicit solution.

4.1.1 Differential Lyapunov Equation

The time varying differential Lyapunov matrix equation is defined by

$$\dot{P}(t) = A^T(t)P(t) + P(t)A(t) + Q(t), \qquad P(t_0) = P_0 = P_0^T \quad (4.1)$$

Its analytical solution has the form

$$P(t) = \Phi(t, t_0)P_0\Phi^T(t, t_0) + \int_{t_0}^{t} \Phi(t, \tau)Q(\tau)\Phi^T(t, \tau)d\tau \quad (4.2)$$

which can be readily verified by differentiating (Kwakernaak and Sivan, 1972).

In (4.2) $\Phi(t, t_0)$ stands for the transition matrix of the time varying system

$$\dot{x}(t) = A(t)x(t)$$

It is important to mention here that no general analytic expression exists for the transition matrix $\Phi(t, t_0)$ when A is time varying (Kwakernaak and Sivan, 1972). In that case a numerical method is the only way to obtain the solution of (4.1). The numerical solution for the time varying Lyapunov differential equation is presented in Section 4.3 following the work of (Subrahmanyam, 1986). For the time invariant case, the transition matrix $\Phi(t - t_0)$ given by

$$\Phi(t - t_0) = e^{A(t-t_0)}$$

is a simple function, so that $P(t)$ as given by (4.2) can be evaluated either analytically or by numerical integration.

Explicit Solution

A direct method for solving the time invariant differential Lyapunov equation is developed in (Rome, 1969) with the restriction that the matrix A is diagonalizable. Making this assumption, we first diagonalize the matrix A through a nonsingular transformation T, that is,

$$TAT^{-1} = D$$

where D is diagonal matrix. Consider the linear system

$$\dot{x}(t) = Ax(t) + \Gamma v(t) \tag{4.3}$$

Define a new state vector $\tilde{x}$ as $\tilde{x} = Tx$ so that the associated state equation becomes

$$\dot{\tilde{x}} = D\tilde{x} + T\Gamma v \tag{4.4}$$

Equation (4.4) represents n decoupled first order equations. Denoting the solution of the Lyapunov equation associated with (4.4) as $\tilde{P}(t)$, the (i, j)-th element of $\tilde{P}(t)$ can be found by the application of (4.2) as

$$\left[\tilde{P}(t)\right]_{ij} = [T\Gamma]_i [T\Gamma]_j^T \frac{(1 - e^{(d_i + d_j)t})}{-((d_i + d_j))} + \left[\tilde{P}(t_0)\right]_{ij} e^{(d_i + d_j)t}$$

where $[T\Gamma]_i$ and $[T\Gamma]_j$ are i-th and j-th rows of $T\Gamma$ and d_i and d_j are i-th and j-th diagonal elements of D. The solution of the original Lyapunov differential equation is obtained by applying the inverse transformation, that is

$$P(t) = T^{-1}\tilde{P}(t)T^{-T}$$

The result of (Rome, 1969) is the only one available in the literature about the methods for explicit solution of the differential Lyapunov equation.

In the following we give conditions that guarantee the existence of a unique solution for both time varying and time invariant differential Lyapunov equation.

Existence Condition

The time varying differential Lyapunov equation has a unique solution under the assumption that the right-hand side of (4.1) is Lipschitz continuous (Coddington and Levinson, 1955). This condition is always satisfied if $A(t)$ and $Q(t)$ are bounded time functions. In addition, for the time invariant differential Lyapunov equation, it is known that *if the matrix A is asymptotically stable, $Q = Q^T \geq 0$, the pair $\left(A, \sqrt{Q}\right)$ is observable, and $P(t_0) = P^T(t_0) > 0$, then there exists a unique positive definite solution for $P(t)$ of the time invariant version of equation (4.1) for all $t \geq t_0$* — see for example (Hmamed, 1990).

4.1.2 Difference Lyapunov Equation

The difference time invariant Lyapunov equation is given by

$$P(t+1) = A^T P(t)A + Q, \qquad P(0) = P_0 = P_0^T \qquad (4.5)$$

Its solution can be obtained by a simple recursion for $t = 1, 2, 3, \ldots,$ which leads to the closed formula

$$P(t) = \left(A^T\right)^t P(0)A^t + \sum_{m=0}^{t-1} \left(A^T\right)^m Q A^m \qquad (4.6)$$

This formula in fact represents the explicit solution of the difference Lyapunov equation (4.5). Note that in this book we use t to represent both continuous-time and discrete-time.

For the time varying difference Lyapunov equation the solution can be similarly derived as in (4.5)-(4.6). Using the notion of the discrete-time transition matrix defined by

$$\Theta(t, t_0) = A(t-1)\ldots A(t_0+1)A(t_0), \quad \Theta(t_0, t_0) = I$$

the solution of the time varying version of (4.5) is given by

$$P(t) = \Theta^T(t, t_0)P(0)\Theta(t, t_0) + \sum_{m=0}^{t-1} \Theta^T(t, m+1)Q(m)\Theta(t, m+1)$$

$$(4.7)$$

The existence of a unique solution of (4.1) is given next.

Existence Condition

It can be seen that the solution of (4.5) as given in (4.6) exists for all bounded matrices A and Q, but it may have infinite escape time. However, like for the differential Lyapunov equation, *assuming that the matrix A is asymptotically stable, $Q = Q^T \geq 0$, the pair $\left(A, \sqrt{Q}\right)$ is observable, and $P_0 = P_0^T > 0$, then there exists a unique positive definite solution to (4.5), $P(t) = P^T(t) > 0$, for all discrete time instants t,* (Hmamed, 1990).

4.2 Bounds of Solution's Attributes

Most of the bounds on the differential and difference Lyapunov equations are described in terms of the solutions of some scalar differential or difference equations.

4.2.1 Eigenvalue Bounds

For the differential and difference Lyapunov equations two theorems giving simultaneous eigenvalue bounds are presented. These theorems are obtained by (Hmamed, 1990). Like in the previous analysis, it has been assumed throughout of this section that the eigenvalues are arranged in decreasing order.

Theorem 4.1 *For the eigenvalues $\lambda_k(P(t))$ of the time invariant differential Lyapunov equation (4.1) the following inequalities hold*

$$(i) \qquad \sum_{i=1}^{k} \lambda_{n-i+1}(P(t)) \geq z_1(t), \quad k = 1, 2, ..., n$$

$$(ii) \qquad \sum_{i=1}^{k} \lambda_i(P(t)) \leq z_2(t), \quad k = 1, 2, ..., n$$

where $z_1(t)$ and $z_2(t)$ are the solutions of the scalar differential equations

$$\dot{z}_1(t) = -2\mu(-A)z_1(t) + \sum_{i=1}^{k} \lambda_{n-i+1}(Q), \quad z_1(0) = \sum_{i=1}^{k} \lambda_{n-i+1}(P_0)$$

$$\dot{z}_2(t) = 2\mu(A)z_2(t) + \sum_{i=1}^{k} \lambda_i(Q), \quad z_2(0) = \sum_{i=1}^{k} \lambda_i(P_0)$$

with $\mu(X) = 0.5\lambda_{max}(X + X^T)$.

$\square$

Proof: The proof of part (i) of Theorem 4.1 follows directly by applying in succession Fan's eigenvalue summation inequality, Amir-Moez inequality, and the inequality from (Mori et al., 1987) to the time invariant version of the solution matrix as given in (4.2). These inequalities are given below.

Fan's Eigenvalue Summation Inequality: (Fan, 1949)

For any $n \times n$ positive semi-definite matrices X and Y the eigenvalues satisfy

$$\sum_{i=1}^{k} \lambda_i(X + Y) \le \sum_{i=1}^{k} \lambda_i(X) + \sum_{i=1}^{k} \lambda_i(Y), \quad k = 1, 2, ..., n \quad (4.8a)$$

with equality for $k = n$. This inequality directly implies another summation inequality

$$\sum_{i=1}^{k} \lambda_{n-i+1}(X + Y) \ge \sum_{i=1}^{k} \lambda_{n-i+1}(X) + \sum_{i=1}^{k} \lambda_{n-i+1}(Y), \quad k = 1, 2, ..., n$$
$$(4.8b)$$

Amir-Moez Inequality: (Amir-Moez, 1956)

For any $n \times n$ positive semi-definite matrices X and Y the following eigenvalue inequalities hold

$$\lambda_{i+j-1}(XY) \le \lambda_i(X)\lambda_j(Y), \quad if \quad i + j \le n + 1 \quad (4.9a)$$
$$\lambda_{i+j-n}(XY) \ge \lambda_i(X)\lambda_j(Y), \quad if \quad i + j \ge n + 1 \quad (4.9b)$$

Eigenvalue Matrix Exponential Inequality: (Mori et al., 1987)

For any real square $n \times n$ matrix X the following is valid

$$\lambda_{min}\left(e^{Xt}e^{X^Tt}\right) \ge e^{-2\mu(-X)t}, \quad \mu = \frac{1}{2}\lambda_{max}(X + X^T), \quad t \ge 0 \quad (4.10)$$

Proof of part (ii) of Theorem 4.1 follows by applying, respectively (4.8a), (4.9a), and Coppel's inequality to (4.2).

Coppel's Inequality: (Coppel, 1965)

For any real square $n \times n$ matrix X the following is satisfied

$$\lambda_{max}\left(e^{Xt}e^{X^T t}\right) \leq e^{2\mu(X)t}, \quad \mu = \frac{1}{2}\lambda_{max}\left(X + X^T\right), \quad t \geq 0 \quad (4.11)$$

Note that inequality (4.10) is derived from Coppel's inequality (Mori et al., 1987).

∎

For the difference Lyapunov equation similar bounds hold. They are summarized in the following theorem.

Theorem 4.2 *For the eigenvalues $\lambda_k(P(t))$ of the difference Lyapunov equation (4.5) the following inequalities hold*

$$(i) \qquad \sum_{i=1}^{k} \lambda_{n-i+1}(P(t+1)) \geq z_1(t+1), \quad k = 1, 2, ..., n$$

$$(ii) \qquad \sum_{i=1}^{k} \lambda_i(P(t+1)) \leq z_2(t+1), \quad k = 1, 2, ..., n$$

$$(iii) \qquad \prod_{i=1}^{k} \lambda_{n-i+1}(P(t+1))^{1/k} \geq z_3(t+1), \quad k = 1, 2, ..., n$$

where $z_i(t+1), i = 1, 2, 3$, are the solutions to the following scalar difference equations

$$z_1(t+1) = \lambda_n\left(AA^T\right)z_1(t) + \sum_{i=1}^{k} \lambda_{n-i+1}(Q)$$

$$z_1(0) = \sum_{i=1}^{k} \lambda_{n-i+1}(P(0))$$

$$z_2(t+1) = \lambda_1\left(AA^T\right)z_2(t) + \sum_{i=1}^{k} \lambda_i(Q), \quad z_2(0) = \sum_{i=1}^{k} \lambda_i(P(0))$$

$$z_3(t+1) = \left\{\prod_{i=1}^{k} \lambda_{n-i+1}\left(AA^T\right)\right\}^{1/k} z_3(t) + \left\{\prod_{i=1}^{k} \lambda_{n-i+1}(Q)\right\}^{1/k}$$

$$z_3(0) = \left\{ \prod_{i=1}^{k} \lambda_{n-i+1}(P(0)) \right\}^{1/k}$$

$\square$

Proof: Apply (4.8b) and (4.9b) to (4.6) in order to get part (i). Using inequalities (4.8a) and (4.9a) applied in the given order to (4.6) implies part (ii). The proof of part (iii) follows from Horn's inequality (2.38a), and the fact that $P \geq 0$ implies that $\left(\prod_{i=1}^{k} \lambda_{n-i+1}(P) \right)^{1/k}$ is a concave function. For more details see (Hmamed, 1990).

∎

The estimates of the upper bounds for any norm of $P(t)$ are also given in (Hmamed, 1990).

4.2.2 Trace Bounds

As with the eigenvalue bounds, the trace bounds are also obtained in terms of the solution of some scalar differential equations. Proceeding on the same lines as in Theorem 4.1, the trace bounds are obtained as given in the following theorem (Mori et al., 1987).

Theorem 4.3 *For the solution of the time invariant differential Lyapunov equation the following trace inequalities are satisfied*

$$z_2(t) \geq \frac{tr(P(t))}{n} \geq z_1(t), \quad t \geq 0$$

where $z_1(t)$ and $z_2(t)$ are the solutions of the scalar differential equations given in Theorem 4.1.

$\square$

Note that $tr(P(t))/n$ represents the arithmetic mean of the eigenvalues of $P(t)$. The above theorem can be also deduced from Theorem 4.1.

Proof: Applying the trace operator to the solution matrix of the time invariant differential Lyapunov equation produces

$$tr\{P(t)\} = tr\left\{ e^{A^T t} P_0 e^{At} \right\} + tr\left\{ \int_0^t e^{A(t-\tau)} Q e^{A^T(t-\tau)} d\tau \right\}$$

Using the trace inequalities from Lemma 2.1 (see also Appendix and Inequality 6), we get

$$tr\{P(t)\} \geq \lambda_{min}\left(e^{A^T t}e^{At}\right)tr\{P_0\}$$
$$+\lambda_{min}(Q)\left\{\int_0^t tr\left\{e^{A(t-\tau)}e^{A^T(t-\tau)}\right\}d\tau\right\} \qquad (4.12)$$

and

$$tr\{P(t)\} \leq \lambda_{max}\left(e^{A^T t}e^{At}\right)tr\{P_0\}$$
$$+\lambda_{max}(Q)\left\{\int_0^t tr\left\{e^{A(t-\tau)}e^{A^T(t-\tau)}\right\}d\tau\right\} \qquad (4.13)$$

Multiplying (4.12) and (4.13) by $\frac{1}{n}$ and applying inequalities (4.10)-(4.11) to (4.12)-(4.13) produces the trace inequalities stated in Theorem 4.3.

∎

4.2.3 Determinant Bounds

Following the same procedure as in the above theorems, the following lower bound is obtained for the determinant of P of the time invariant differential Lyapunov equation, (Mori et al., 1986).

Theorem 4.4 *The determinant of the solution of the time invariant differential Lyapunov equation satisfies*

$$|P(t)|^{1/n} \geq z(t), \quad t \geq 0$$

where $z(t)$ is the solution to

$$\dot{z}(t) = \frac{2}{n}tr(A)z(t) + |Q|^{1/n}, \quad z(0) = |P(0)|^{1/n}$$

□

Note that $|P(t)|^{1/n}$ represents the geometric mean of the eigenvalues of $P(t)$.

The proof of this theorem is lengthy and the reader interested in it is referred to the original paper (Mori et al., 1986). Note that in the limit as $t \to \infty$ Theorem 4.4 produces the lower bound which coincides with the determinant bound obtained in Section 2.2.3 for the determinant of the solution of the algebraic Lyapunov equation.

4.2.4 Solution Bounds

Very interesting result for the solution bounds is obtained in (Geromel and Bernussou, 1979). This result is stated in the following lemma.

Lemma 4.1 *The solution of the time invariant differential Lyapunov equation satisfies the following inequalities*

$$g(\mu_1, t)Q \leq P(t) \leq g(\mu_2, t)Q, \quad \forall t \geq 0$$

where

$$\mu_1 = \max_x \left\{ \frac{x^T \left(-A^T Q - QA \right) x}{x^T Q x} \right\}$$

$$\mu_2 = \min_x \left\{ \frac{x^T \left(-A^T Q - QA \right) x}{x^T Q x} \right\}$$

and

$$g(\zeta, t) = \int_0^t e^{-\zeta\tau} d\tau$$

□

Complete proof of this lemma can be found in (Geromel and Bernussou, 1979).

This interesting result at steady state produces also the bounds for the solution of the corresponding continuous-time algebraic Lyapunov equation as

$$\frac{1}{\mu_1}Q \leq P \leq \frac{1}{\mu_2}Q, \quad \text{provided } \mu_2 > 0 \qquad (4.14)$$

4.3 Numerical Solutions

Numerical solutions for the differential Lyapunov equation appear only in the literature, and therefore they are presented here. The difference Lyapunov equation can be numerically solved as a simple recursion.

The first algorithm to be described here is given in (Davison, 1975), which is in fact an adaptation of Algorithm 2.4 for the algebraic Lyapunov equation. As in Algorithm 2.4, it is assumed that A is a stable matrix. The algorithm uses the fact that the unique solution of the time invariant version of (4.1) is given by

$$P(t) = e^{A^T t}(P(0) - E)e^{At} + E$$

where E is the solution of the algebraic matrix Lyapunov equation

$$A^T E + EA + Q = 0$$

Here E can be calculated through Algorithm 2.6 (Bartels and Stewart, 1972) as suggested by (Barraud, 1977), but it is preferable to use Algorithm 2.4, because the value of C (which is an approximation for e^{Ah} where h is a step size) is needed for the next step, see Algorithm 2.4. Therefore, the use of Algorithm 2.4 to compute E will eliminate the computation of C for the next step. The algorithm is as follows.

Algorithm 4.1:
1. Use Algorithm 2.4 to compute E.
2. For $j = 0, 1, 2, \ldots$ calculate

$$P(hj) = (C^T)^j (P(0) - E)C^j + E$$

where C is obtained in Step 2 of Algorithm 2.4, and h is a small positive constant.

$$\Delta$$

The accuracy of the presented algorithm is $O(h^5)$, (Davison, 1973, 1975). As with Algorithm 2.4, this algorithm is also stable and converges for all small $h > 0$. The algorithm requires $8n^2$ memory locations and $4n^3(k + 2)$ multiplications, where k is the number of iterations required for Step 1. Note that if A is a real symmetric matrix then it can be diagonalized and the calculation of C, which is in fact an approximation of e^{Ah}, can be computed rather easily (Lacoss and Shakal, 1976). Algorithms similar to Davison's algorithm for numerical solution of the differential Lyapunov equation have been developed by (Barraud, 1977; Hoskins et al., 1977). It is interesting to point out that all of these algorithms (Davidson, 1975; Barraud, 1977; Hoskins et al., 1977) use the same method for the approximation of the matrix exponential

(Padé approximation), but they differ in the methods used for solving the corresponding algebraic Lyapunov equation.

Direct integration of (4.1) without solving the algebraic Lyapunov equation, based on a version of Van Loan's procedure for computing integrals involving matrix exponentials (Van Loan, 1978) is presented in (Serbin and Serbin, 1980).

The next method described in (Subrahmanyam, 1986) is a modification of the Runge-Kutta method. This method treats the general case when the matrix A is time varying. The basic idea of the algorithm is to first approximate $P(t)A(t) + Q(t)$ on the interval $[t_i, t_{i+1}]$, by using Runge-Kutta like approximation (hence $A(t)$ and $P(t)A(t) + Q(t)$ are constant in the said interval), and then to use the truncated Taylor series. Suppose that $A(t)$ is approximated by a constant matrix A and $P(t)A(t) + Q(t)$ is approximated by a constant matrix R in the given interval, then the new differential equation to be solved is

$$\dot{S}(t) = A^T S(t) + R \tag{4.15}$$

The algorithm discussed in (Subrahmanyam, 1986) consists of two passes, each of which consists of an approximation of $P(t)A(t) + Q(t)$ and the solution of an equation of the form (4.15) obtained by using truncated Taylor series. The development is as follows.

Consider the interval $[t_i, t_{i+1}]$. On both passes, A and Q are assumed to be constant at their values at $t_{i+1/2} = t_i + h/2$, where $h > 0$ is the step size, that is

$$A = A(t_{i+1/2}), \quad Q = Q(t_{i+1/2})$$

PASS 1: The computed value of $P(t_i)$ is known. To differentiate it from the actual value of $P(t_i)$, we use the notation $P^c(t_i)$ for a computed value of $P(t)$ at $t = t_i$. Approximating $P(t)A(t) + Q(t)$ by

$$R_1 = P^c(t_i)A(t_{i+1/2}) + Q(t_{i+1/2})$$

we get an initial value problem

$$\dot{S}_1(t) = A^T S_1(t) + R_1, \qquad S_1(t_i) = P^c(t_i) \tag{4.16}$$

The Taylor series expansion of $S_1(t)$ is

$$S_1(t_{i+1}) = \sum_{k=0}^{\infty} S_1^{(k)}(t_i)\frac{h^k}{k!}$$

From (4.16) we have

$$S_1^{(k)}(t_i) = (A^T)^k S_1(t_i) + A^{k-1} R_1 = (A^T)^k P^c(t_i) + (A^T)^{k-1} R_1$$

Using last two expressions we obtain

$$S_1(t_{i+1}) = \sum_{k=0}^{\infty}(A^T)^k P^c(t_i)\frac{h^k}{k!} + \sum_{k=0}^{\infty}(A^T)^k R_1\frac{h^{k+1}}{(k+1)!}$$

It is shown in (Subrahmanyam, 1986) that in order to obtain local truncation error of order of $O(h^3)$, we need to use only first two terms in the first sum, and first term in the second sum. Therefore, truncating the sums in the last expressions, we obtain an approximation

$$S_1(t_{i+1}) = P^c(t_i) + hA^T P^c(t_i) + hR_1$$

PASS 2: Let

$$\hat{P} = \frac{S_1(t_i) + S_1(t_{i+1})}{2}, \quad R_2 = \hat{P}A(t_{i+1/2}) + Q(t_{i+1/2})$$

then we have the initial value problem

$$\dot{S}_2(t) = A^T S_2(t) + R_2, \quad S_2(t_i) = P^c(t_i)$$

Proceeding as in pass 1

$$S_2(t_{i+1}) = \sum_{k=0}^{\infty}(A^T)^k P^c(t_i)\frac{h^k}{k!} + \sum_{k=0}^{\infty}(A^T)^k R_2\frac{h^{k+1}}{(k+1)!}$$

in order to obtain the local truncation error of order of $O(h^3)$, we need to use first three terms in the first sum, and first two terms in the second sum, that is

$$S_2(t_{i+1}) = P^c(t_i)(I + hA^T + \frac{h^2}{2}(A^T)^2) + R_2(h + \frac{h^2}{2}A^T)$$

Finally, we set $P^c(t_{i+1}) = S_2(t_{i+1})$.

Summarizing, we have the following algorithm.

Algorithm 4.2:

On the interval $[t_i, t_{i+1}]$, $P^c(t_i)$ is known

1. Let $A = A(t_{i+1/2})$, $Q = Q(t_{i+1/2})$.
2. Compute

$$R_1 = P^c(t_i)A + Q$$

$$S_1 = (I + hA^T)P^c(t_i) + hR_1$$

$$\hat{P} = \frac{1}{2}(S_1 + P^c(t_i))$$

$$R_2 = \hat{P}A + Q$$

$$S_2(t_{i+1}) = P^c(t_i)(I + hA^T + \frac{h^2}{2}(A^T)^2) + R_2(h + \frac{h^2}{2}A^T)$$

3. Set $S_2(t_{i+1}) = P^c(t_{i+1}) = $ the completed value of $P(t)$ at $t = t_{i+1}$.

$$\Delta$$

4.4 Singularly Perturbed and Weakly Coupled Systems[1]

In this section, we obtain the reduced-order equations for the solution of differential and difference Lyapunov equations for linear singularly perturbed and weakly coupled systems (Kokotovic and Khalil, 1986; Gajic et al., 1990; Gajic and Shen, 1993). These reduced-order equations are much easier to solve than the full-order differential and difference Lyapunov equations. More about singularly perturbed and weakly coupled linear systems can be found at the beginning of Chapter 5.

The obtained reduced-order differential equations are either differential Lyapunov equations or they have similar structures to the Lyapunov equation. The techniques for obtaining such equations include i) partitioning of the full-order Lyapunov equation according to the special structures of the matrices A, P, and Q, and ii) block diagonalizing

---

[1]   Some parts from this section are reprinted, with permission from the publisher, from the book *Parallel Algorithms for Optimal Control of Large Scale Linear Systems*, by Z. Gajic and X. Shen, Springer Verlag, London, 1993.

the matrix A by some nonsingular transformation. In this section, we will consider the so-called regulator type differential Lyapunov equation (Kwakernaak and Sivan, 1972) given by

$$\dot{P} = A^T P + PA + Q, \qquad P(t_0) = P_0 \qquad (4.17)$$

The so-called filter type differential Lyapunov equation is obtained by taking the transpose of the regulator Lyapunov equation.

In the following we assume that all coefficient matrices are time invariant. For notational convenience, we have dropped the dependence of all coefficient matrices on perturbation parameter ϵ. The derivatives are taken with respect to time t.

In the second part of this section, we perform the similar order-reduction technique for the difference Lyapunov equations of singularly perturbed and weakly coupled systems.

4.4.1 Singularly Perturbed Differential Lyapunov Equation

The matrices A, P, and Q of the singularly perturbed differential Lyapunov equation have the following structures (Kokotovic and Khalil, 1986)

$$A = \begin{bmatrix} A_1 & A_2 \\ \frac{1}{\epsilon}A_3 & \frac{1}{\epsilon}A_4 \end{bmatrix}, \quad P = \begin{bmatrix} P_1 & \epsilon P_2 \\ \epsilon P_2^T & \epsilon P_3 \end{bmatrix}, \quad Q = \begin{bmatrix} Q_1 & Q_2 \\ Q_2^T & Q_3 \end{bmatrix} \qquad (4.18)$$

where $A_1, P_1, Q_1 \in \Re^{n_1 \times n_1}$, $A_4, P_3, Q_3 \in \Re^{n_2 \times n_2}$ and ϵ is a small positive singular perturbation parameter. Scaling introduced in (4.18) simplifies the form of reduced-order equations. In addition, it is assumed that all matrices are continuous function of ϵ.

Substituting (4.18) into (4.17), we obtain the following reduced-order differential equations

$$\begin{aligned} \dot{P}_1 &= A_1^T P_1 + P_1 A_1 + A_3^T P_2^T + P_2 A_3 + Q_1 \\ \epsilon \dot{P}_2 &= \epsilon A_1^T P_2 + P_1 A_2 + A_3^T P_3 + P_2 A_4 + Q_2 \qquad (4.19) \\ \epsilon \dot{P}_3 &= \epsilon A_2^T P_2 + \epsilon P_2^T A_2 + A_4^T P_3 + P_3 A_4 + Q_3 \end{aligned}$$

These equations are singularly perturbed differential equations due to the fact that the derivatives of P_2 and P_3 are multiplied by the small parameter ϵ. It is known that equations (4.19) are numerically ill-conditioned (Miranker, 1981; Kokotovic and Khalil, 1986). Variables P_2 and P_3 are changing quickly (fast variables) and the variable P_1 is slow.

Equations in (4.19) can be decoupled by using the approach adopted in (Gajic et al., 1990), where the matrix A *is block diagonalized* by some nonsingular transformation, and then, the reduced-order equations are derived. The resulting equations become simpler than those in (4.19). Several different transformations are introduced in the literature, among which the most widely used is obtained by (Chang, 1972), for which a new version is presented in (Qureshi and Gajic, 1992). The Chang transformation is given by

$$\mathbf{T_1} = \begin{bmatrix} I & \epsilon H_1 \\ -L_1 & I - \epsilon L_1 H_1 \end{bmatrix} \Rightarrow \mathbf{T_1}^{-1} A \mathbf{T_1} = \begin{bmatrix} A_s & 0 \\ 0 & \frac{1}{\epsilon} A_f \end{bmatrix} \quad (4.20)$$

where matrices L_1 and H_1 satisfy the following algebraic matrix equations

$$A_4 L_1 - \epsilon L_1 A_1 + \epsilon L_1 A_2 L_1 - A_3 = 0$$
$$\epsilon(A_1 - A_2 L_1)H_1 - H_1(A_4 + \epsilon L_1 A_2) + A_2 = 0 \quad (4.21)$$

Note that for $\epsilon = 0$ these equations have unique solutions $L_1^{(0)} = A_4^{-1} A_3$ and $H_1^{(0)} = A_2 A_4^{-1}$ under the assumption that *the matrix A_4 is nonsingular*, which is the standard assumption in the theory of singular perturbations (Kokotovic and Khalil, 1986). Also, by the implicit function theorem the nonsingularity of A_4 implies the existence of the unique solutions for L_1 and H_1 in the ϵ-neighborhoods of $L_1^{(0)}$ and $H_1^{(0)}$. Efficient numerical algorithms for solving equations (4.21) can be found in (Grodt and Gajic, 1988), where the fixed-point iterations and the Newton method are presented.

Multiplying (4.17) from the left-hand side by $\mathbf{T_1^T}$ and from the right-hand side by $\mathbf{T_1}$, we obtain

$$\mathbf{T_1^T} \dot{P} \mathbf{T_1} = \mathbf{T_1^T} A^T P \mathbf{T_1} + \mathbf{T_1^T} P A \mathbf{T_1} + \mathbf{T_1^T} Q \mathbf{T_1} \quad (4.22)$$

which can be written as

$$\dot{K} = a^T K + K a + q, \quad K(t_0) = K_0 = \mathbf{T_1^T} P_0 \mathbf{T_1} \quad (4.23)$$

where, by using (4.21), we have

$$a = \mathbf{T}_1^{-1} A \mathbf{T}_1 = \begin{bmatrix} A_s & 0 \\ 0 & \frac{1}{\epsilon} A_f \end{bmatrix}, \quad q = \mathbf{T}_1^{\mathbf{T}} Q \mathbf{T}_1, \quad K = \mathbf{T}_1^{\mathbf{T}} P \mathbf{T}_1 \quad (4.24)$$

with

$$A_s = A_1 - A_2 L_1, \quad A_f = A_4 + \epsilon L_1 A_2$$

The matrices K and q have the following structures

$$K = \begin{bmatrix} K_1 & \epsilon K_2 \\ \epsilon K_2^T & \epsilon K_3 \end{bmatrix}, \quad q = \begin{bmatrix} q_1 & q_2 \\ q_2^T & q_3 \end{bmatrix} \quad (4.25)$$

Partitioning (4.23) according to (4.24)-(4.25) produces a system of completely decoupled reduced-order differential equations

$$\dot{K}_1 = A_s^T K_1 + K_1 A_s + q_1$$
$$\epsilon \dot{K}_2 = \epsilon A_s^T K_2 + K_2 A_f + q_2 \quad (4.26)$$
$$\epsilon \dot{K}_3 = A_f^T K_3 + K_3 A_f + q_3$$

with initial conditions defined in (4.23). Note that all of the reduced-order equations obtained in (4.26) are either Lyapunov or Lyapunov-like equations. Equations (4.26) represent the desired decomposition. Since dimensions of these equations are considerably reduced they can be solved with less computational efforts. In addition, the slow part of the solution (K_1) and the fast solution's parts (K_2 and K_3) are now completely decoupled so that the numerical ill-conditioning is removed. However, the variables (K_2 and K_3) are represented by stiff differential equations so that the solution of these equations is still computationally involved (Miranker, 1981). The original solution $P(t)$ can be obtained from (4.24) as $P(t) = \mathbf{T}_1^{-\mathbf{T}} K(t) \mathbf{T}_1^{-1}$.

4.4.1.1 Case Study: A DC Motor

Consider a singularly perturbed DC motor from (Yackel and Kokotovic, 1972). Its state space form is given by

$$\begin{bmatrix} \dot{x}_1 \\ \dot{x}_2 \end{bmatrix} = \begin{bmatrix} 0 & \frac{D}{G} \\ -\frac{C}{L} & -\frac{R}{L} \end{bmatrix} \begin{bmatrix} x_1 \\ x_2 \end{bmatrix} + \begin{bmatrix} 0 \\ \frac{1}{L} \end{bmatrix} u$$

with the following numerical data: $R = 7.9\Omega$, $L = 0.0136H$, $C = D = 0.0246Vs/rad$, and $G = 1.32 \times 10^{-6}kgm^2$. This system can be put in the singularly perturbed form by using the change of variables $x_1 = z_1, x_2 = \epsilon z_2$ with $\epsilon = 0.001$ leading to

$$\begin{bmatrix} \dot{z}_1 \\ \epsilon \dot{z}_2 \end{bmatrix} = \begin{bmatrix} 0 & 18.6363 \\ -1.8088 & -0.5809 \end{bmatrix} \begin{bmatrix} x_1 \\ x_2 \end{bmatrix} + \begin{bmatrix} 0 \\ 73.529 \end{bmatrix} u$$

Consider the Lyapunov differential equation of this system with $Q = I_2$. Since $\epsilon = 0.001$ is quite small the corresponding problem is quite stiff for numerical integration. However, by using the presented methodology the solution can be obtained in terms of the reduced-order completely decoupled slow and fast differential equations. From (4.21) we get the values for the matrices L_1 and H_1 and the required transformation as follows

$$L_1 = 3.5088, \ H_1 = -41.4030 \ \Rightarrow \ T_1 = \begin{bmatrix} 1.0000 & -0.0414 \\ -3.5088 & 1.1453 \end{bmatrix}$$

The slow and fast subsystem matrices and the matrix q are obtained from (4.24) as

$$A_s = -65.3903, \ A_f = -0.5155, \ q = \begin{bmatrix} 13.3114 & -4.0599 \\ -4.0599 & 1.3134 \end{bmatrix}$$

so that the required differential equations are

$$\dot{K}_1 = -130.7806 K_1 + 13.3114$$
$$\epsilon \dot{K}_2 = -0.5809 K_2 - 4.0599$$
$$\epsilon \dot{K}_3 = -1.031 K_3 + 1.3134$$

Assuming that the initial condition for the original Lyapunov equation is specified we can get the initial conditions for the transformed differential equations by using (4.23). The obtained pure-slow and pure-fast differential equations can be solved analytically easily so that the solution of the differential singularly perturbed Lyapunov equation is obtained through the reverse procedure, that is, by using $P(t) = T_1^{-T} K(t) T_1^{-1}$.

4.4.2 Weakly Coupled Differential Lyapunov Equation

In this section, we proceed on the same lines as with singularly perturbed systems considered Section 4.4.1. The weakly coupled systems have the following structures for matrices A, P, and Q, (Gajic et al., 1990)

$$A = \begin{bmatrix} A_1 & \epsilon A_2 \\ \epsilon A_3 & A_4 \end{bmatrix}, \quad P = \begin{bmatrix} P_1 & \epsilon P_2 \\ \epsilon P_2^T & P_3 \end{bmatrix}, \quad Q = \begin{bmatrix} Q_1 & \epsilon Q_2 \\ \epsilon Q_2^T & Q_3 \end{bmatrix} \quad (4.27)$$

Substituting (4.27) in (4.17) we obtain the following partitioned equations

$$\begin{aligned} \dot{P}_1 &= A_1^T P_1 + P_1 A_1 + \epsilon^2 \left(A_3^T P_2^T + A_3 P_2 \right) + Q_1 \\ \dot{P}_2 &= A_1^T P_2 + A_3^T P_3 + P_1 A_2 + P_2 A_4 + Q_2 \\ \dot{P}_3 &= A_4^T P_3 + P_3 A_4 + \epsilon^2 \left(A_2^T P_2 + P_2^T A_2 \right) + Q_3 \end{aligned} \quad (4.28)$$

These equations are coupled differential Lyapunov equations.

If we apply the transformation approach *to block diagonalize matrix* A, we will obtain the decoupled Lyapunov differential equations. The corresponding nonsingular transformation for weakly coupled systems has been introduced in (Gajic and Shen, 1989), and is given by

$$\mathbf{T_2} = \begin{bmatrix} I & -\epsilon L_2 \\ \epsilon H_2 & I - \epsilon^2 H_2 L_2 \end{bmatrix}, \quad \mathbf{T_2}^{-1} = \begin{bmatrix} I - \epsilon^2 L_2 H_2 & \epsilon L_2 \\ -\epsilon H_2 & I \end{bmatrix}$$

$$a = \mathbf{T_2} A \mathbf{T_2^{-1}} = \begin{bmatrix} A_{10} & 0 \\ 0 & A_{40} \end{bmatrix} \quad (4.29)$$

where matrices L_2 and H_2 satisfy the following algebraic equations

$$\begin{aligned} A_1 L_2 - L_2 A_4 + A_2 - \epsilon^2 L_2 A_3 L_2 &= 0 \\ H_2 \left(A_1 - \epsilon^2 L_2 A_3 \right) - \left(A_4 + \epsilon^2 A_3 L_2 \right) H_2 + A_3 &= 0 \end{aligned} \quad (4.30)$$

These equations have unique solutions for sufficiently small values of perturbation parameter ϵ under the assumption that *the matrices A_1 and A_4 have no eigenvalues in common* (Gajic and Shen, 1989). A recursive approach for solving (4.30) as a sequence of linear algebraic equations is also given in (Gajic and Shen, 1989). Another version of this transformation that also decouples the transformation equations for L_2 and H_2 is derived in (Qureshi, 1992).

Multiplying the full-order Lyapunov differential equation (4.17) from left by $\mathbf{T_2^{-T}}$ and from right by $\mathbf{T_2^{-1}}$, we obtain the following Lyapunov differential equation

$$\dot{K} = a^T K + Ka + q, \quad K(t_0) = K_0 \qquad (4.31)$$

where

$$a = \mathbf{T_2} A \mathbf{T_2^{-1}} = \begin{bmatrix} A_{10} & 0 \\ 0 & A_{40} \end{bmatrix}, \quad q = \mathbf{T_2^{-T}} Q \mathbf{T_2^{-1}}, \quad K = \mathbf{T_2^{-T}} P \mathbf{T_2^{-1}} \qquad (4.32)$$

with

$$A_{10} = A_1 - \epsilon^2 L_2 A_3, \quad A_{40} = A_4 + \epsilon^2 A_3 L_2$$

The following partitioning and scaling are used for matrices K and q

$$K = \begin{bmatrix} K_1 & \epsilon K_2 \\ \epsilon K_2^T & K_3 \end{bmatrix}, \quad q = \begin{bmatrix} q_1 & \epsilon q_2 \\ \epsilon q_2^T & q_3 \end{bmatrix} \qquad (4.33)$$

Partitioning (4.31) according to (4.32)-(4.33) reveals completely decoupled reduced-order differential equations

$$\begin{aligned}
\dot{K}_1 &= K_1 A_{10} + A_{10}^T K_1 + q_1 \\
\dot{K}_2 &= K_2 A_{40} + A_{10}^T K_2 + q_2 \\
\dot{K}_3 &= K_3 A_{40} + A_{40}^T K_3 + q_3
\end{aligned} \qquad (4.34)$$

with initial conditions defined in (4.31)-(4.32). Having obtained $K_i, i = 1, 2, 3$, from (4.34), we can get the solution of the Lyapunov differential equation in the original coordinates as

$$P = \mathbf{T_2^T} K \mathbf{T_2} \qquad (4.35)$$

The decomposition techniques for the *difference* singularly perturbed and weakly coupled Lyapunov equations will be presented in the next subsections.

4.4.3 Singularly Perturbed Difference Lyapunov Equation

Singularly perturbed difference Lyapunov equation (in its fast time scale representation (Gajic and Shen, 1993)) is given by

$$P(t+1) = A^T P(t) A + Q, \qquad P(t_0) = P_0 = P_0^T$$

where

$$A = \begin{bmatrix} I + \epsilon A_1 & \epsilon A_2 \\ A_3 & A_4 \end{bmatrix}, \qquad Q = \begin{bmatrix} Q_1 & Q_2 \\ Q_2^T & Q_3 \end{bmatrix}, \qquad P = \begin{bmatrix} \frac{1}{\epsilon} P_1 & P_2 \\ P_2^T & P_3 \end{bmatrix}$$

$$(4.36)$$

The following transformation can be used in order to decompose singularly perturbed difference Lyapunov equation (Gajic and Shen, 1993, page 140)

$$\mathbf{T_3} = \begin{bmatrix} I - \epsilon H_3 L_3 & -\epsilon H_3 \\ L_3 & I \end{bmatrix}, \qquad \mathbf{T_3^{-1}} = \begin{bmatrix} I & \epsilon H_3 \\ -L_3 & I - \epsilon L_3 H_3 \end{bmatrix} \qquad (4.37)$$

where

$$\begin{aligned} (I - A_4) L_3 + \epsilon L_3 A_1 + A_3 - \epsilon L_3 A_2 L_3 &= 0 \\ H_3 (I - A_4 - \epsilon L_3 A_2) + \epsilon (A_1 - A_2 L_3) H_3 + A_2 &= 0 \end{aligned} \qquad (4.38)$$

leading to

$$\mathbf{T_3} A \mathbf{T_3^{-1}} = \begin{bmatrix} a_s & 0 \\ 0 & a_f \end{bmatrix} = \begin{bmatrix} I + \epsilon A_1 - \epsilon A_2 L_3 & 0 \\ 0 & A_4 + \epsilon L_3 A_2 \end{bmatrix} \qquad (4.39)$$

Note that the existence of unique solutions for equations (4.38) for small values of ϵ requires that $det(I - A_4) \neq 0$ (Kokotovic and Khalil, 1986; Gajic and Shen, 1993). The change of variables

$$P(t) = \mathbf{T_3^T} K(t) \mathbf{T_3} = \mathbf{T_3^T} \begin{bmatrix} \frac{1}{\epsilon} K_1(t) & K_2(t) \\ K_2^T(t) & K_3(t) \end{bmatrix} \mathbf{T_3}$$

$$(4.40)$$

$$q = \mathbf{T_3^{-T}} Q \mathbf{T_3^{-1}} = \begin{bmatrix} q_1 & q_2 \\ q_2^T & q_3 \end{bmatrix}$$

produces in the new coordinates completely decoupled Lyapunov (Lyapunov-like) difference equations

$$
\begin{aligned}
K_1(t+1) &= a_s^T K_1(t)a_s + \epsilon q_1 \\
K_2(t+1) &= a_s^T K_2(t)a_f + q_2 \\
K_3(t+1) &= a_f^T K_3(t)a_f + q_3
\end{aligned}
\tag{4.41}
$$

with a_s, a_f easily obtained from (4.39), and initial conditions calculated from (4.40).

Note that the full-order difference singularly perturbed Lyapunov equation is numerically ill-conditioned and that the decoupled equations (4.41) are well-defined.

4.4.4 Weakly Coupled Difference Lyapunov Equation

The coefficient matrices for the weakly coupled difference Lyapunov equation have exactly the same form as the corresponding ones of the differential weakly coupled Lyapunov equation, that is, they are given by (4.27). Using the same transformation like for the continuous-time domain, (4.29)-(4.30), the difference weakly coupled Lyapunov equation given by

$$
\begin{aligned}
&\begin{bmatrix} P_1(t+1) & \epsilon P_2(t+1) \\ \epsilon P_2^T(t+1) & P_3(t+1) \end{bmatrix} = \begin{bmatrix} Q_1 & \epsilon Q_2 \\ \epsilon Q_2^T & Q_3 \end{bmatrix} \\
&+ \begin{bmatrix} A_1 & \epsilon A_2 \\ \epsilon A_3 & A_4 \end{bmatrix}^T \begin{bmatrix} P_1(t) & \epsilon P_2(t) \\ \epsilon P_2^T(t) & P_3(t) \end{bmatrix} \begin{bmatrix} A_1 & \epsilon A_2 \\ \epsilon A_3 & A_4 \end{bmatrix}
\end{aligned}
\tag{4.42}
$$

is converted into decoupled reduced-order forms similar to (4.41), and the original solution is recovered by using (4.40) with the transformation matrix obtained from (4.29)-(4.30).

The order-reduction decomposition techniques presented in this section can be generalized to include Lyapunov differential and difference equations containing several subsystems.

4.5 Coupled Differential Lyapunov Equations

Coupled differential Lyapunov equations are considered in the work of (Jodar and Mariton, 1987). They appear in the control problems of jump linear systems (Mariton, 1990). The following set of differential coupled Lyapunov equations is studied

$$\dot{P}_i = A_i^T(t)P_i + P_i A_i(t) + Q_i(t) + \sum_{j=1}^{N} D_{ij}(t)P_j \tag{4.43}$$

$$i, j = 1, 2, ..., N; \qquad P_j(t_0) = P_{j0}$$

where all coefficient matrices are continuous time functions. Using the Kronecker product representation the above equation can be represented by

$$\dot{\widehat{P}}(t) = M(t)\widehat{P}(t), \qquad \widehat{P}(t_0) = \widehat{P}_0 \tag{4.44}$$

where

$$M_{ii}(t) = I \otimes [A_i(t) + D_{ii}(t)] + A_i(t) \otimes I$$
$$M_{ij}(t) = I \otimes D_{ij}(t), \quad i, j = 1, 2, ..., N; \quad i \neq j \tag{4.45}$$

and vectors $\widehat{P}(t), \widehat{Q}(t), \widehat{P}(t_0)$ are obtained by stacking columns of matrices $P(t), Q(t), P(t_0)$ into the corresponding vectors, that is

$$\widehat{P}(t) = vec(P), \; P(t) = \begin{bmatrix} P_1 \\ P_2 \\ \vdots \\ P_N \end{bmatrix}, \; \widehat{P}(t_0) = vec(P_0), \; P_0 = \begin{bmatrix} P_{10} \\ P_{20} \\ \vdots \\ P_{N0} \end{bmatrix}$$

$$\widehat{Q}(t) = vec(Q), \quad Q(t) = \begin{bmatrix} Q_1 \\ Q_2 \\ \vdots \\ Q_N \end{bmatrix}$$

$$\tag{4.46}$$

Let $\Phi(t, t_0)$ represent the transition matrix of the time varying system (4.44), then the solution of the coupled differential Lyapunov equations (4.43) is given by (Jodar and Mariton, 1987)

$$\widehat{P}(t) = \Phi(t, t_0) \left[\widehat{P}(t_0) + \int_{t_0}^{t} \Phi(t_0, \tau)\widehat{Q}(\tau)d\tau \right] \tag{4.47}$$

Two-point boundary value problem for the coupled differential Lyapunov equations is also studied in (Jodar and Mariton, 1987).

4.6 Summary

The solution of the differential Lyapunov matrix equation is given in terms of the system transition matrix by formula (4.2). The transition matrix can be, in general, obtained by numerical integration. A direct method for solving the differential Lyapunov equation is described in the case when the coefficient matrix A is diagonalizable. For the difference Lyapunov equation the solution can be obtained by simple iterative calculations.

The bounds of the main attributes of the solutions of the differential and difference Lyapunov equations are also described. They are mostly given in terms of the solutions of some scalar differential and difference equations. These bounds include eigenvalue bounds, trace bounds, determinant bounds, and solution bounds.

The numerical solutions are described only for the differential Lyapunov equation. The first algorithm assumes that the coefficient matrix is time invariant and stable. The second algorithm is more general and is valid for any time varying coefficient matrix. It uses the Runge-Kutta-like approximation on the interval $[t_i, t_{i+1}]$, and the truncated Taylor series. The algorithm has two passes each of which consists of an approximation and construction of the truncated Taylor series. Local truncation error is also discussed in order to obtain an arbitrary degree of accuracy.

In Section 4.5, reduced-order solutions for differential and difference Lyapunov matrix equations for two special classes of systems with small parameters are presented. These systems are singularly perturbed and weakly coupled linear systems. It has been shown that the Lyapunov equations for these systems can be decomposed into three lower order Lyapunov/Lyapunov-like equations. Last section of this chapter shows how to convert a set of coupled differential Lyapunov equations into an initial value problem.

4.7 References

1. Amir-Moez, A., "Extreme properties of a Hermitian transformation and singular values of the sum and product of linear transformation," *Duke Math., J.*, vol.23, 463–476, 1956.

2. Barraud, A., "A new numerical solution of $\dot{X} = A_1 X + X A_2 + D$, $X(0) = C$," *IEEE Trans. Automatic Control*, vol.22, 976–977, 1977.

3. Bartels, R. and G. Stewart, "Algorithm 432, solution of the matrix equation $AX + XB = C$," *Comm. Ass. Computer Machinery*, vol.15, 820-826, 1972.

4. Chang, K., "Singular perturbations of a general boundary value problem," *SIAM J. Math. Anal.*, vol.3, 520–526, 1972.

5. Coddington, E. and N. Levinson, *Theory of Ordinary Differential Equations*, McGraw-Hill, New York, 1955.

6. Coppel, W., *Stability and Asymptotic Behavior of Differential Equations*, Heath, Boston, 1965.

7. Davison, E., "An algorithm for computer simulation of very large dynamic systems," *Automatica*, vol.9, 665–675, 1973.

8. Davison, E. "The numerical solution of $\dot{X} = A_1 X + X A_2 + D$, $X(0) = C$," *IEEE Trans. Automatic Control*, vol.20, 566–567, 1975.

9. Davison, E. and F. Man, "The numerical solution of $A^T Q + Q A = -C$," *IEEE Trans. Automatic Control*, vol.13, 448–449, 1968.

10. Fan, K., "On the theorem of Weyl concerning eigenvalues of linear transformations," *Proc. National Academy of Science, USA*, vol.35, 652–655, 1949.

11. Gajic, Z. and X. Shen, "Decoupling transformation for weakly coupled linear systems," *Int. J. Control*, vol.50, 1517–1523, 1989.

12. Gajic, Z., D. Petkovski, and X. Shen, *Singularly Perturbed and Weakly Coupled Linear Control Systems: A Recursive Approach*, Lecture Notes in Control and Information Sciences, Springer Verlag, Berlin, 1990.

13. Gajic, Z. and X. Shen, *Parallel Algorithms for Optimal Control of Linear Large Scale Systems*, Springer Verlag, London, 1993.

14. Geromel, J. and J. Bernussou, "On bounds of Lyapunov matrix equation," *IEEE Trans. Automatic Control*, vol.24, 482–483, 1979.

15. Grodt, T. and Z. Gajic, "The recursive reduced order numerical solution of the singularly perturbed matrix differential Riccati equation," *IEEE Trans. Automatic Control*, vol.33, 751–754, 1988.

16. Hmamed, A., "Differential and difference Lyapunov equations: simultaneous eigenvalue bounds," *Int. J. Systems Science*, vol.21, 1335–1344, 1990.

17. Hoskins, W., D. Meek, and D. Walton, "The numerical solution of $\dot{X} = A_1 X + X A_2 + D$, $X(0) = C$," *IEEE Trans. Automatic Control*, vol.22, 881–882, 1977.

18. Jodar, L. and M. Mariton, "Explicit solutions for a system of coupled Lyapunov differential matrix equations," *Proc. Edinburgh Math. Soc.*, vol.30, 427–434, 1987.

19. Kokotovic, P. and H. Khalil, *Singular Perturbations in Systems and Control*, IEEE Press, 1986.

20. Kwakernaak, H. and R. Sivan, *Linear Optimal Control Systems*, Wiley, 1972.

21. Lacoss, R. and A. Shakal, "More $A_1E + EA_2 = -D$ and $\dot{X} = A_1X + XA_2 + D$, $X(0) = C$," *IEEE Trans. Automatic Control*, vol.21, 405–406, 1976.

22. Mariton, M., *Jump Linear Systems in Automatic Control*, Marcel Dekker, New York — Basel, 1990.

23. Miranker, W., *Numerical Methods for Stiff Equations and Singular Perturbation Problems*, Reidel Publishing Company, Dordrecht, Holland, 1981.

24. Mori, T., N. Fukuma, and M. Kuwahara, "On the Lyapunov matrix differential equation," *IEEE Trans. Automatic Control*, vol.31, 868–869, 1986.

25. Mori, T., N. Fukuma, and M. Kuwahara, "Bounds in the Lyapunov matrix differential equations," *IEEE Trans. Automatic Control*, vol.32, 55-57, 1987.

26. Qureshi, M., *Parallel Algorithms for Discrete Singularly Perturbed and Weakly Coupled Filtering and Control Problems*, Doctoral Dissertation, Rutgers University, 1992.

27. Qureshi, M. and Z. Gajic, "A new version of the Chang's transformation," *IEEE Trans. Automatic Control*, vol.37, 800–801, 1992.

28. Rome, H., "A direct solution to the linear variance equation of a time invariant linear system," *IEEE Trans. Automatic Control*, vol.14, 592–593, 1969.

29. Serbin, S. and C. Serbin, "A time-stepping procedure for $\dot{X} = A_1X + XA_2 + D$, $X(0) = C$," *IEEE Trans. Automatic Control*, vol.25, 1138–1141, 1980.

30. Subrahmanyam, M., "On a numerical method of solving the Lyapunov and Sylvester equations," *Int. J. Control*, vol.43, 433–439, 1986.

31. Van Loan, C., "Computing integrals involving the matrix exponentials," *IEEE Trans. Automatic Control*, vol.23, 395–404, 1978.

32. Yackel, R. and P. Kokotovic, "A boundary layer method for the matrix Riccati equation," *IEEE Trans. Automatic Control*, vol.18, 17–24, 1973.

Chapter Five

Algebraic Lyapunov Equations with Small Parameters[1]

There are two special classes of systems where some small parameters appear pertaining to the special behavior of such systems. These two classes are known as singularly perturbed and weakly coupled systems. In this chapter, we study the *algebraic* Lyapunov equations for both singularly perturbed and weakly coupled systems. We derive corresponding recursive algorithms for solving these equations in the general cases when the problem matrices are functions of small perturbation parameters. The numerical decompositions have been achieved, so that only low-order systems are involved in algebraic computations. The introduced recursive methods can be implemented as synchronous parallel algorithms (Bertsekas and Tsitsiklis, 1989, 1991).

Both continuous-time and discrete-time versions of the algebraic Lyapunov equations are studied. It is shown that the singular perturbation recursive methods converge with the rate of convergence of $O(\epsilon)$, whereas the recursive methods for weakly coupled linear systems converge faster, that is, with the rate of convergence of $O(\epsilon^2)$, where ϵ is a small perturbation parameter.

---

[1]   This chapter is reprinted in parts, with permission from the publisher, from the book *Parallel Algorithms for Optimal Control of Large Scale Linear Systems*, by Z. Gajic and X. Shen, Springer Verlag, London, 1993.

The corresponding *differential and difference* Lyapunov equations of singularly perturbed and weakly coupled systems have been presented in the previous chapter in Section 4.4.

The continuous-time linear singularly perturbed systems are described by the following state equations (Kokotovic et al., 1986)

$$\dot{x}_1(t) = A_1 x_1(t) + A_2 x_2(t)$$
$$\epsilon \dot{x}_2(t) = A_3 x_1(t) + A_4 x_2(t) \tag{5.1}$$

where $x_1(t) \in \Re^{n_1}$ and $x_2(t) \in \Re^{n_2}$ are state vectors and matrices A_1, A_2, A_3, A_4 are of compatible dimensions and they can be constant or time varying. A small *positive* parameter ϵ indicates separation of state variables into slow, x_1, and fast, x_2, variables. Note that for $\epsilon = 0, \dot{x}_2(t)$ becomes infinite, hence singularity is observed.

Modeling a system by a singular perturbation model may not be an easy task. Such models are well suited for those systems that exhibit multiple time scale behavior, that is, the system may contain components with large and small time constants, or some parasitic elements may be present (which are normally ignored to obtain reduced-order system), or the system may contain high feedback gains. Many real world examples are given in (Kokotovic et al., 1986; Kokotovic and Khalil, 1986; Gajic and Shen, 1993) where such modeling is discussed. In all such cases the singular perturbation models ease the analysis by removing the inherent ill-conditioning in the system and depicting the different time scale behavior separately. It also increases the accuracy of the solution by taking all the elements into account, which may be neglected in some cases.

The continuous-time linear weakly coupled systems are described by the following state space differential equations (Kokotovic et al., 1969; Gajic et al., 1990)

$$\dot{x}_1(t) = A_1 x_1(t) + \epsilon A_2 x_2(t)$$
$$\dot{x}_2(t) = \epsilon A_3 x_1(t) + A_4 x_2(t) \tag{5.2}$$

where $x_1(t) \in R^{n_1}$ and $x_2(t) \in R^{n_2}$ are state vectors and matrices A_1, A_2, A_3, and A_4 are of compatible dimensions. A small parameter ϵ of *arbitrary sign* couples the states $x_1(t)$ and $x_2(t)$.

The singularly perturbed linear system in discrete-time is given by (Litkouhi and Khalil, 1984, 1985; Gajic and Shen, 1993)

$$x_1(t+1) = (I + \epsilon A_1)x_1(t) + \epsilon A_2 x_2(t)$$
$$x_2(t+1) = A_3 x_1(t) + A_4 x_2(t) \tag{5.3}$$

while the weakly coupled linear discrete system is described by (Shen, 1990; Shen and Gajic 1990a, 1990b; Shen et al., 1991)

$$x_1(t+1) = A_1 x_1(t) + \epsilon A_2 x_2(t)$$
$$x_2(t+1) = \epsilon A_3 x_1(t) + A_4 x_2(t) \tag{5.4}$$

where the dimensions of $x_1(t), x_2(t), A_1, A_2, A_3,$ and A_4 are the same as for continuous-time systems.

For both classes of small parameter systems we will consider the general full-order system

$$\dot{x}(t) = Ax(t) \tag{5.5}$$

in the continuous-time case, and

$$x(t+1) = Ax(t) \tag{5.6}$$

in the discrete-time case with the system matrix A having special structures. The structure of the matrix A for continuous singularly perturbed systems is

$$A = \begin{bmatrix} A_1 & A_2 \\ \frac{1}{\epsilon}A_3 & \frac{1}{\epsilon}A_4 \end{bmatrix} \tag{5.7}$$

while for discrete singularly perturbed systems it has the form

$$A = \begin{bmatrix} I + \epsilon A_1 & \epsilon A_2 \\ A_3 & A_4 \end{bmatrix} \tag{5.8}$$

For weakly coupled systems the matrix A has the same structure for both continuous-time and discrete-time cases and is given by

$$A = \begin{bmatrix} A_1 & \epsilon A_2 \\ \epsilon A_3 & A_4 \end{bmatrix} \tag{5.9}$$

With such structures for the system matrix A, in the next sections, the algebraic subsystem Lyapunov equations are obtained and the recursive reduced-order algorithms are presented for numerical solutions of the corresponding algebraic Lyapunov equations. Two case studies are included in order to demonstrate efficiency of the presented algorithms, which can be easily implemented in MATLAB.

5.1 Singularly Perturbed Continuous Lyapunov Equation

In this section, we study a dual form of the algebraic Lyapunov equation that represents a variance equation of a linear system driven by white noise

$$\dot{x}(t) = A(\epsilon)x(t) + G(\epsilon)w(t) \tag{5.10}$$

where w is a zero-mean Gaussian white noise process with a unity intensity matrix. The algebraic Lyapunov equation corresponding to (5.10) is given by

$$P(\epsilon)A^T(\epsilon) + A(\epsilon)P(\epsilon) + G(\epsilon)G^T(\epsilon) = 0 \tag{5.11}$$

According to the theory of singular perturbations (Kokotovic et al., 1986; Kokotovic and Khalil, 1986), the following partition and scaling of the problem matrices are introduced

$$P(\epsilon) = \begin{bmatrix} P_1(\epsilon) & P_2(\epsilon) \\ P_2^T(\epsilon) & \frac{1}{\epsilon}P_3(\epsilon) \end{bmatrix}, \quad G(\epsilon) = \begin{bmatrix} G_1(\epsilon) \\ \frac{1}{\epsilon}G_2(\epsilon) \end{bmatrix} \tag{5.12}$$

Newly defined matrices are of dimensions $P_1 \in \Re^{n_1 \times n_1}$, $P_3 \in \Re^{n_2 \times n_2}$, $G_i \in \Re^{n_i \times m}$, $i = 1, 2$, with $n_1 + n_2 = n$. It is assumed that all matrices are continuous functions of ϵ.

The partitioned form of the Lyapunov equation defined in (5.11) subject to (5.7) and (5.12) is

$$A_1(\epsilon)P_1(\epsilon) + P_1(\epsilon)A_1^T(\epsilon) + A_2(\epsilon)P_2^T(\epsilon) + P_2(\epsilon)A_2^T(\epsilon) \\ + G_1(\epsilon)G_1^T(\epsilon) = 0$$

$$P_2(\epsilon)A_4^T(\epsilon) + \epsilon A_1(\epsilon)P_2(\epsilon) + P_1(\epsilon)A_3^T(\epsilon) + A_2(\epsilon)P_3(\epsilon) \\ + G_1(\epsilon)G_2^T(\epsilon) = 0 \tag{5.13}$$

$$P_3(\epsilon)A_4^T(\epsilon) + A_4(\epsilon)P_3(\epsilon) + \epsilon A_3(\epsilon)P_2(\epsilon) + \epsilon P_2^T(\epsilon)A_3^T(\epsilon) \\ + G_2(\epsilon)G_2^T(\epsilon) = 0$$

Define an $O(\epsilon)$ perturbation of (5.13) as

$$A_1(\epsilon)\mathbf{P_1}(\epsilon) + \mathbf{P_1}(\epsilon)A_1^T(\epsilon) + A_2(\epsilon)\mathbf{P_2^T}(\epsilon) + \mathbf{P_2}(\epsilon)A_2^T(\epsilon)$$
$$+G_1(\epsilon)G_1^T(\epsilon) = 0$$

$$\mathbf{P_2}(\epsilon)A_4^T + \mathbf{P_1}(\epsilon)A_3^T(\epsilon) + A_2(\epsilon)\mathbf{P_3}(\epsilon) + G_1(\epsilon)G_2^T(\epsilon) = 0 \tag{5.14}$$

$$\mathbf{P_3}(\epsilon)A_4^T(\epsilon) + A_4(\epsilon)\mathbf{P_3}(\epsilon) + G_2(\epsilon)G_2^T(\epsilon) = 0$$

Note that we did not set $\epsilon = 0$ in $A_i, i = 1, 2, 3, 4$, and $G_j, j = 1, 2$. In the rest of the chapter we will assume that all matrices are functions of ϵ. However, the explicit dependence on ϵ of the problem matrices will be omitted in order to simplify notation. Solution of (5.14) is in fact given in terms of two low-order algebraic Lyapunov equations

$$A_0\mathbf{P_1} + \mathbf{P_1}A_0^T + G_0G_0^T = 0$$
$$A_4\mathbf{P_3} + \mathbf{P_3}A_4^T + G_2G_2^T = 0 \tag{5.15}$$

and

$$\mathbf{P_2} = -\left(\mathbf{P_1}A_3^T + A_2\mathbf{P_3} + G_1G_2^T\right)A_4^{-T} \tag{5.16}$$

where

$$A_0 = A_1 - A_2A_4^{-1}A_3, \quad G_0 = G_1 - A_2A_4^{-1}G_2 \tag{5.17}$$

Unique solutions of (5.15)-(5.16) exist under the following assumption.

Assumption 5.1: Matrices $A_0(\epsilon)$ and $A_4(\epsilon)$ are asymptotically stable.

$$\Delta$$

This is a standard assumption in the theory of singular perturbations (Kokotovic et al., 1986; Kokotovic and Khalil, 1986). Defining the approximation errors as

$$P_1 = \mathbf{P_1} + \epsilon E_1, \quad P_2 = \mathbf{P_2} + \epsilon E_2, \quad P_3 = \mathbf{P_3} + \epsilon E_3 \tag{5.18}$$

and subtracting (5.15)-(5.16) from (5.13), we get the error equations (after some algebra) in the form

$$A_0E_1 + E_1A_0^T = A_0[\mathbf{P_2} + \epsilon E_2]A_4^{-T}A_2^T$$
$$+A_2A_4^{-1}[\mathbf{P_2} + \epsilon E_2]^TA_0^T$$

$$A_4 E_3 + E_3 A_4^T = -A_3[\mathbf{P_2} + \epsilon E_2] - [\mathbf{P_2} + \epsilon E_2]^T A_3^T \qquad (5.19)$$

$$A_2 E_3 + E_1 A_3^T + E_2 A_4^T + A_1[\mathbf{P_2} + \epsilon E_2] = 0$$

These equations have very nice forms since the unknown quantity E_2 in equations for E_1 and E_3 is multiplied by the small parameter ϵ. This fact suggests the following reduced-order parallel algorithm for solving (5.19).

Algorithm 5.1: (Gajic et al., 1989)

$$A_0 E_1^{(i+1)} + E_1^{(i+1)} A_0^T = A_0 \left[\mathbf{P_2} + \epsilon E_2^{(i)} \right] A_4^{-T} A_2^T$$
$$+ A_2 A_4^{-1} \left[\mathbf{P_2} + \epsilon E_2^{(i)} \right]^T A_0^T$$

$$A_4 E_3^{(i+1)} + E_3^{(i+1)} A_4^T = -A_3 \left[\mathbf{P_2} + \epsilon E_2^{(i)} \right] - \left[\mathbf{P_2} + \epsilon E_2^{(i)} \right]^T A_3^T$$

$$E_2^{(i+1)} = -\left\{ E_1^{(i+1)} A_3^T + A_2 E_3^{(i+1)} + A_1 \left[\mathbf{P_2} + \epsilon E_2^{(i)} \right] \right\} A_4^{-T}$$
$$(5.20)$$

with starting point $E_2^{(0)} = 0$.

$$\Delta$$

Theorem 5.1 (Gajic et al., 1989). *Under the stability assumptions imposed on $A_0(\epsilon)$ and $A_4(\epsilon)$, the algorithm (5.20) converges to the exact solution E with the rate of convergence of $O(\epsilon)$, and thus, the required solution P can be obtained with the accuracy of $O(\epsilon^i)$ from*

$$P_j^{(i)} = \mathbf{P_j} + \epsilon E_j + O(\epsilon^i), \quad j = 1, 2, 3; \quad i = 1, 2, \dots \qquad (5.21)$$

$$\square$$

Proof: Using the stability property imposed in Assumption 5.1, it is easy to show that (5.20) is a contraction mapping, (Luenberger, 1969), that is

$$\left\| E_j^{(i)} - E_j \right\| = O(\epsilon^i), \quad j = 1, 2, 3; \quad i = 1, 2, \dots \qquad (5.22)$$

Note that (5.22) is valid in the case when the last equation of (5.20) is in the form

$$E_2^{(i+1)} = \left\{ E_1^{(i)} A_3^T + A_2 E_3^{(i)} + A_1 \left[\mathbf{P_2} + \epsilon E_2^{(i)} \right] \right\} A_4^{-T} \qquad (5.23)$$

with $E_1^{(0)} = 0$ and $E_3^{(0)} = 0$. Thus, the algorithm (5.20) is convergent. Using $E_j^{(\infty)}$, $j = 1, 2, 3$, in (5.20) and comparing it to (5.19), imply that the algorithm (5.20) converges to the unique solution of (5.19), which proves the theorem.

■

It is important to note that in the proposed method we do not need to expand $A_i(\epsilon), i = 1, 2, 3, 4$, into power series, and we do not require stability of $A_0(0)$ and $A_4(0)$, which makes the important features of the presented method. However, $det A_0(\epsilon)$ and $det A_4(\epsilon)$ must be $O(1)$. Assumption 5.1 is more natural and less binding than the stability assumption of $A_0(0)$ and $A_4(0)$, (Kokotovic et al., 1986; Kokotovic and Khalil, 1986). Namely, the singularly perturbed structure of the system is the consequence of a strict inequality $\epsilon > 0$ (small positive parameter). The stability requirement imposed on $A_0(0)$ and $A_4(0)$ is based on the continuation argument, but it can not be indefinitely exploited.

It is known that the power series expansion method leads to two reduced-order Lyapunov equations similar to those in (5.20) - they are of the same order, but the number of terms on the right-hand side of these equations for the power series expansion method is growing very quickly with the increase in the required accuracy. It can be seen from (5.20) that for the fixed point method the number of terms on the right-hand side is constant. The number of matrix multiplications required to form right-hand sides of Lyapunov equations, corresponding to the fast variables, that is $E_3^{(i+1)}$, for the accuracy of $O(\epsilon^i)$, is given in Table 5.1.

i	1	2	3	4	5	6
Fixed Point	1	1	1	1	1	1
Power Series	3	6	9	12	15	18

Table 5.1: Required number of matrix multiplications

This table shows a very strong support for the proposed reduced-order fixed-point method. Finally, the important advantage of the presented fixed-point algorithm is in its parallel and distributed structure.

Note that the proposed parallel Algorithm 5.1 is a synchronous one (Bert-sekas and Tsitsiklis, 1989, 1991).

MATLAB Program for Algorithm 5.1

Algorithm 5.1 can be easily coded in MATLAB as given below.

```
% input a1, a2, a3, a4, g1, g2, eps, n1, n2, N
% zeroth-order approximation (5.15)-(5.16)
a0=a1-a2*inv(a4)*a3;
g0=g1-a2*inv(a4)*g2;
p10=lyap(a0,g0*g0');
p30=lyap(a4,g2*g2');
p20=-(a2*p30+p10*a3'+g1*g2')*inv(a4');
p0=[p10 p20;p20' p30/eps]
% initialization
e1i=zeros(n1,n1);
e2i=zeros(n1,n2);
e3i=zeros(n2,n2);
% fixed point iterations for the errors (5.20)
for i=0:N
p1i=p10+eps*e1i;
p2i=p20+eps*e2i;
p3i=p30+eps*e3i;
q1i=a0*p2i*inv(a4')*a2'+a2*inv(a4)*p2i*a0';
e1i=lyap(a0,-q1i);
q3i=a3*p2i+p2i'*a3';
e3i=lyap(a4,q3i);
e2i=-(a2*e3i+e1i*a3'+a1*p2i)*inv(a4');
pi=[p1i p2i;p2i' p3i/eps]
% error at i-th iteration
a=[a1 a2;a3/eps a4/eps];
g=[g1;g2/eps];
p=lyap(a,g*g')
err=p-pi
pause
end
```

A special class of singularly perturbed algebraic Lyapunov equations is efficiently studied in (Derbel et al., 1994) by using the Taylor series expansion.

5.2 Weakly Coupled Continuous Lyapunov Equation

The weakly coupled linear continuous-time system is defined in (5.2). In this section, we develop the recursive parallel algorithm for solving the algebraic Lyapunov equation of weakly coupled systems. The algebraic Lyapunov equation for weakly coupled systems ('regulator type') is given by

$$A^T(\epsilon)P(\epsilon) + P(\epsilon)A(\epsilon) + Q(\epsilon) = 0 \qquad (5.24)$$

Due to the block structure of matrices A and Q, the required solution P, and the matrix Q are properly scaled as follows

$$P(\epsilon) = \begin{bmatrix} P_1(\epsilon) & \epsilon P_2(\epsilon) \\ \epsilon P_2^T(\epsilon) & P_3(\epsilon) \end{bmatrix}, \quad Q(\epsilon) = \begin{bmatrix} Q_1(\epsilon) & \epsilon Q_2(\epsilon) \\ \epsilon Q_2^T(\epsilon) & Q_3(\epsilon) \end{bmatrix} \qquad (5.25)$$

The partitioned form of (5.24) produces

$$P_1 A_1 + A_1^T P_1 + Q_1 + \epsilon^2 \left(P_2 A_3 + A_3^T P_2^T \right) = 0$$
$$P_3 A_4 + A_4^T P_3 + Q_3 + \epsilon^2 \left(P_2^T A_2 + A_2^T P_2 \right) = 0 \qquad (5.26)$$
$$P_2 A_4 + A_1^T P_2 + P_1 A_2 + A_3^T P_3 + Q_2 = 0$$

We define an $O(\epsilon^2)$ approximation of (5.26) as

$$\mathbf{P_1} A_1 + A_1^T \mathbf{P_1} + Q_1 = 0$$
$$\mathbf{P_3} A_4 + A_4^T \mathbf{P_3} + Q_3 = 0 \qquad (5.27)$$
$$\mathbf{P_2} A_4 + A_1^T \mathbf{P_2} = -\mathbf{P_1} A_2 - A_3^T \mathbf{P_3} - Q_2$$

Note that we did not set $\epsilon = 0$ in $A_i, i = 1, 2, 3, 4$, and $Q_j, j = 1, 2, 3$, so that $\mathbf{P}_j, j = 1, 2, 3$, are functions of ϵ. The required solution of (5.27) exists under the following assumption.

Assumption 5.2: Matrices $A_1(\epsilon)$ and $A_4(\epsilon)$ are asymptotically stable.

$\triangle$

Defining errors as

$$P_j = \mathbf{P_j} + \epsilon^2 E_j, \quad j = 1, 2, 3 \tag{5.28}$$

and subtracting (5.27) from (5.26), we obtain the error equations

$$E_1 A_1 + A_1^T E_1 + \mathbf{P_2} A_3 + A_3^T \mathbf{P_2^T} + \epsilon^2 \left(E_2 A_3 + A_3^T E_2^T \right) = 0$$

$$E_2 A_4 + A_1^T E_2 + E_1 A_2 + A_3^T E_3 = 0 \tag{5.29}$$

$$E_3 A_4 + A_4^T E_3 + \mathbf{P_2^T} A_2 + A_2^T \mathbf{P_2} + \epsilon^2 \left(A_2^T E_2 + E_2^T A_2 \right) = 0$$

We propose the following algorithm, having reduced-order and parallel structure, for solving (5.29)

Algorithm 5.2:

$$
\begin{aligned}
E_1^{(i+1)} A_1 + A_1^T E_1^{(i+1)} &= -\left\{ P_2^{(i)} A_3 + A_3^T P_2^{(i)^T} \right\} \\
E_3^{(i+1)} A_4 + A_4^T E_3^{(i+1)} &= -\left\{ P_2^{(i)^T} A_2 + A_2^T P_2^{(i)} \right\} \\
E_2^{(i+1)} A_4 + A_1^T E_2^{(i+1)} &= -\left\{ E_1^{(i+1)} A_2 + A_3^T E_3^{(i+1)} \right\}
\end{aligned}
\tag{5.30}
$$

$$i = 1, 2, \dots$$

with the starting point $E_2^{(0)} = 0$ and

$$P_j^{(i)} = \mathbf{P_j} + \epsilon^2 E_j^{(i)}, \quad j = 1, 2, 3; \ i = 0, 1, 2, \dots \tag{5.31}$$

$$\triangle$$

Using the same arguments like in Section 5.1, the following theorem can be established.

Theorem 5.2 (*Shen et al., 1991*). *Under stability assumptions imposed on the matrices $A_1(\epsilon)$ and $A_4(\epsilon)$, the algorithm (5.30) converges to the exact solution E with the rate of convergence of $O(\epsilon^2)$, and thus, the required solution P can be obtained with the accuracy of $O(\epsilon^{2i})$ from (5.31), that is*

$$P_j = P_j^{(i)} + O\left(\epsilon^{2i} \right), \quad j = 1, 2, 3; \ i = 1, 2, \dots \tag{5.32}$$

$$\square$$

5.3 Singularly Perturbed Discrete Systems

The linear singularly perturbed discrete systems have been studied recently in different setups by many researchers. Two main structures of singularly perturbed linear discrete systems are considered: the fast time scale version (Litkouhi and Khalil, 1984, 1985; Butuzov and Vasileva, 1971; Hoppensteadt and Miranker, 1977; Blankenship, 1981; Oloomi and Sawan, 1987; Khorasani and Azimi-Sadjadi, 1987) and the slow time scale version (Phillips, 1980; Naidu and Rao, 1985). Discrete-time models of the singularly perturbed linear systems, similar to (Phillips, 1980; Naidu and Rao, 1985), were studied by Mahmoud and his coworkers also (Mahmoud et al., 1986). Since the slow time scale version presupposes the asymptotic stability of the fast modes, it seems that in the design procedure of stabilizing feedback controllers, the fast time scale version is much more appropriate (Litkouhi and Khalil, 1985). In this section, we adopt the fast time scale version of the singularly perturbed discrete systems.

The proposed method for solving the algebraic Lyapunov equation of discrete singularly perturbed systems produces the reduced-order near-optimal solution, up to an arbitrary degree of accuracy, that is of $O(\epsilon^k)$, where ϵ is a small positive perturbation parameter. In addition, it reduces the size of required computations. The real world example, an F-8 aircraft, demonstrates the efficiency of the presented method.

5.3.1 Parallel Algorithm for Discrete Algebraic Lyapunov Equation

A discrete-time invariant linear, input free, system

$$x(t+1) = Ax(t) \tag{5.33}$$

is asymptotically stable if and only if the algebraic discrete Lyapunov equation

$$A^T P A - P = -Q \tag{5.34}$$

has a positive definite solution for some positive definite symmetric matrix Q. The transpose of the Lyapunov equation (5.34) represents a variance equation of a linear stochastic discrete-time system driven by a Gaussian zero-mean white noise process $w(t)$ with the intensity matrix Q

$$x(t+1) = Ax(t) + w(t) \tag{5.35}$$

Consider the algebraic discrete Lyapunov equation (5.34) of the singularly perturbed linear discrete-time system represented in the fast time scale version by the corresponding matrices (Litkouhi and Khalil, 1985)

$$A = \begin{bmatrix} I + \epsilon A_1 & \epsilon A_2 \\ A_3 & A_4 \end{bmatrix}, \; P = \begin{bmatrix} \frac{1}{\epsilon}P_1 & P_2 \\ P_2^T & P_3 \end{bmatrix}, \; Q = \begin{bmatrix} Q_1 & Q_2 \\ Q_2^T & Q_3 \end{bmatrix} \tag{5.36}$$

$A_i, i = 1, 2, 3, 4$, and $Q_j, j = 1, 2, 3$, are assumed to be continuous functions of ϵ. Matrices P_1 and P_3 are of dimensions $n_1 \times n_1$ and $n_2 \times n_2$, respectively. Remaining matrices are of compatible dimensions.

The partitioned form of (5.34) subject to (5.36) is given by

$$\begin{aligned} & P_1 A_1 + A_1^T P_1 + P_2 A_3 + A_3^T P_2^T + A_3^T P_3 A_3 + Q_1 \\ & + \epsilon \left(A_1^T P_1 A_1 + A_1^T P_2 A_3 + A_3^T P_2^T A_1 \right) = 0 \end{aligned} \tag{5.37}$$

$$P_1 A_2 + P_2 A_4 + A_3^T P_3 A_4 - P_2 + Q_2 + \epsilon \left(A_1^T P_2 A_4 + A_3^T P_2^T A_2 \right) = 0 \tag{5.38}$$

$$A_4^T P_3 A_4 - P_3 + Q_3 + \epsilon \left(A_2^T P_1 A_2 + A_2^T P_2 A_4 + A_4^T P_2^T A_2 \right) = 0 \tag{5.39}$$

Define an $O(\epsilon)$ perturbation of (5.37)-(5.39) by

$$\mathbf{P_1} A_1 + A_1^T \mathbf{P_1} + \mathbf{P_2} A_3 + A_3^T \mathbf{P_2^T} + A_3^T \mathbf{P_3} A_3 + Q_1 = 0 \tag{5.40}$$

$$\mathbf{P_1} A_2 + \mathbf{P_2} A_4 + A_3^T \mathbf{P_3} A_4 - \mathbf{P_2} + Q_2 = 0 \tag{5.41}$$

$$A_4^T \mathbf{P_3} A_4 - \mathbf{P_3} + Q_3 = 0 \tag{5.42}$$

Note that we did not set $\epsilon = 0$ in $A_i, i = 1, 2, 3, 4$, and $Q_j, j = 1, 2, 3$. From equation (5.41) the matrix $\mathbf{P_2}$ can be expressed in terms of $\mathbf{P_1}$ and $\mathbf{P_3}$ as

$$\mathbf{P_2} = L_1 + \mathbf{P_1} L_2 \tag{5.43}$$

where

$$L_1 = \left(A_3^T \mathbf{P_3} A_4 + Q_2 \right)(I - A_4)^{-1}, \; L_2 = A_2(I - A_4)^{-1} \tag{5.44}$$

The invertibility of the matrix $(I - A_4)$ follows from the stability assumption, that is, $|\lambda(A_4)| < 1$.

After doing some algebra we get

$$\mathbf{P_1}\mathbf{A} + \mathbf{A}^T\mathbf{P_1} + \mathbf{Q} = 0 \tag{5.45}$$

where

$$\begin{aligned}
\mathbf{A} &= A_1 + L_2 A_3 = A_1 + A_2(I - A_4)^{-1}A_3 \\
\mathbf{Q} &= L_1 A_3 + A_3^T L_1^T + A_3^T \mathbf{P_3} A_3 - Q_1
\end{aligned} \tag{5.46}$$

Thus, we can get solutions for $\mathbf{P_1}, \mathbf{P_2}$, and $\mathbf{P_3}$ by solving one low-order *continuous-time* algebraic Lyapunov equation (5.45) and one low-order *discrete-time* Lyapunov equation (5.42). It is assumed that $\mathbf{A}$ and A_4 are stable matrices in the corresponding time domains, so that unique solutions of (5.42) and (5.45) exist. These are standard assumptions in the theory of singularly perturbed linear discrete-time systems (Litkouhi and Khalil, 1985).

Assumption 5.3: The matrix $\mathbf{A}$ is asymptotically stable in the continuous-time domain and the matrix A_4 is asymptotically stable in the discrete-time domain.

$$\Delta$$

Define errors as

$$P_1 = \mathbf{P_1} + \epsilon E_1, \quad P_2 = \mathbf{P_2} + \epsilon E_2, \quad P_3 = \mathbf{P_3} + \epsilon E_3 \tag{5.47}$$

Subtracting (5.40)-(5.42) from (5.37)-(5.39) and doing some algebra, the following set of error equations is obtained

$$\begin{aligned}
E_1 \mathbf{A} + \mathbf{A}^T E_1 &= -H A_3 - A_3^T H^T - A_3^T E_3 A_3 - A_3^T E_3 A_4 \\
&\quad - A_4^T E_3 A_3 - A_1^T P_1 A_1 - A_3^T P_2^T A_1 - A_1^T P_2 A_3
\end{aligned} \tag{5.48a}$$

$$A_4^T E_3 A_4 - E_3 = -\left(A_2^T P_1 A_2 + A_2^T P_2 A_4 + A_4^T P_2^T A_2\right) \tag{5.48b}$$

$$E_2 = E_1 L_2 + H + A_3^T E_3 A_4 \tag{5.48c}$$

where

$$H = \left(A_3^T E_3 A_4 + A_1^T P_1 A_2 + A_3^T P_2^T A_2 + A_1^T P_2 A_4\right)(I - A_4)^{-1} \tag{5.49}$$

The solution of (5.48) of the given accuracy will produce the same accuracy for the solution of the considered Lyapunov equation (5.34). The proposed parallel algorithm for the numerical solution of (5.48) is as follows:

Algorithm 5.3: (Qureshi et al., 1992).

$$E_1^{(i+1)} A + \mathbf{A}^T E_1^{(i+1)} = -H^{(i)} A_3 - A_3^T H^{(i)^T} - A_3^T E_3^{(i)} A_3$$

$$- A_3^T E_3^{(i)} A_4 - A_4^T E_3^{(i)} A_3 - A_1^T P_1^{(i)} A_1 - A_3^T P_2^{(i)^T} A_1 - A_1^T P_2^{(i)} A_3 \tag{5.50a}$$

$$A_4^T E_3^{(i+1)} A_4 - E_3^{(i+1)} = -\left(A_2^T P_1^{(i)} A_2 + A_2^T P_2^{(i)} A_4 + A_4^T P_2^{(i)^T} A_2 \right) \tag{5.50b}$$

$$E_2^{(i+1)} = E_1^{(i+1)} L_2 + H^{(i)} + A_3^T E_3^{(i+1)} A_4 \tag{5.50c}$$

with starting points $E_1^{(0)} = E_2^{(0)} = E_3^{(0)} = 0$ and

$$P_j^{(i)} = \mathbf{P_j} + \epsilon E_j^{(i)}, \quad j = 1, 2, 3; \quad i = 0, 1, 2, \dots \tag{5.51}$$

$$\Delta$$

Main feature of Algorithm 5.3 is given in the following theorem.

Theorem 5.3 *(Qureshi et al., 1992). Based on the stability assumptions imposed on* $\mathbf{A}$ *and* A_4, *the algorithm (5.50)-(5.51) converges to the exact solutions for* $E_j, j = 1, 2, 3$, *with the rate of convergence of* $O(\epsilon)$.

$$\square$$

The proof of this theorem is similar to the proof of the corresponding algorithm for the continuous-time algebraic singularly perturbed Lyapunov equation studied in Section 5.1. It uses the bilinear transformation from (Power, 1967) to transform the discrete-time Lyapunov equation into the continuous one and then follows the ideas of Section 5.1. A direct, discrete-time domain proof is given in (Qureshi et al., 1992) in the context of the output feedback control problem of discrete-time systems.

A numerical example is solved in the next subsection for a linearized model of an F-8 aircraft in order to demonstrate the efficiency of the presented method.

5.3.2 Case Study: An F-8 Aircraft

The problem matrices are given by (Litkouhi, 1983)

$$A = 10^{-3} \begin{bmatrix} 998.51 & -8.044 & -0.1089 & -0.0187 \\ 0.15659 & 1000.0 & -0.7623 & 3.2272 \\ -213.94 & 0.8808 & 897.21 & 92.826 \\ 110.17 & -0.3782 & -445.56 & 929.68 \end{bmatrix}$$

$$Q = diag\{0.1, \quad 0.1, \quad 0.1, \quad 0.1\}$$

The small parameter ϵ is equal to 0.03333. The simulation results, obtained by using the software package L-A-S (Bingulac and Vanlandingham, 1993), are presented in Table 5.2.

i	$P_1^{(i)}$	$P_2^{(i)}$	$P_3^{(i)}$
0	7.26070 0.17779 0.17779 8.02970	-1.4252 -0.8117 -7.3907 1.2959	2.10690 -0.25231 -0.25231 0.54983
1	8.15990 0.20340 0.20340 8.95420	-1.6239 -0.8927 -8.2950 1.4750	2.33520 -0.29196 -0.29196 0.55729
2	8.27660 0.20656 0.20656 9.07410	-1.6494 -0.9033 -8.4104 1.4976	2.36500 -0.29714 -0.29714 0.55884
3	8.28940 0.20696 0.20696 9.08960	-1.6526 -0.9047 -8.4253 1.5006	2.36880 -0.29781 -0.29781 0.55884
4	8.29130 0.20701 0.20701 9.09160	-1.6531 -0.9048 -8.4272 1.5010	2.36930 -0.29790 -0.29790 0.55887
5	8.29160 0.20702 0.20702 9.09180	-1.6531 -0.9049 -8.4275 1.5010	2.36930 -0.29791 -0.29791 0.55887

Table 5.2: Solution of the singularly perturbed
discrete algebraic Lyapunov equation

The full-order problem is numerically very ill-conditioned due to relatively small value for ϵ. However, the well-conditioned reduced-order equations from Algorithm 5.3 have produced the desired solution after only 5 iterations.

5.4 Recursive Methods for Weakly Coupled Discrete Systems

The weakly coupled linear discrete systems have been studied by (Shen, 1990; Shen and Gajic, 1990a, 1990b; Shen et al., 1991). A parallel reduced-order algorithm for solving the discrete Lyapunov equation of weakly coupled systems is derived and demonstrated on a catalytic cracker model example. As before, the algorithm for the Lyapunov equation is implemented as synchronous one. Its implementation as asynchronous parallel algorithm is under investigation.

5.4.1 Parallel Algorithm for Discrete Algebraic Lyapunov Equation

Consider the algebraic discrete Lyapunov equation (5.34). In the case of the weakly coupled linear discrete system the corresponding matrices are partitioned as

$$A = \begin{bmatrix} A_1 & \epsilon A_2 \\ \epsilon A_3 & A_4 \end{bmatrix}, \; P = \begin{bmatrix} P_1 & \epsilon P_2 \\ \epsilon P_2^T & P_3 \end{bmatrix}, \; Q = \begin{bmatrix} Q_1 & \epsilon Q_2 \\ \epsilon Q_2^T & Q_3 \end{bmatrix} \quad (5.52)$$

Matrices A_i, $i = 1, 2, 3, 4$, and Q_j, $j = 1, 2, 3$, are assumed to be continuous functions of ϵ. Matrices P_1 and P_3 are of dimensions $n_1 \times n_1$ and $n_2 \times n_2$, respectively. Remaining matrices are of compatible dimensions.

The partitioned form of (5.34) subject to (5.52) is

$$A_1^T P_1 A_1 - P_1 + Q_1 + \epsilon^2 \left(A_1^T P_2 A_3 + A_3^T P_2^T A_1 + A_3^T P_3 A_3 \right) = 0 \quad (5.53)$$

$$A_1^T P_1 A_2 - P_2 + Q_2 + A_1^T P_2 A_4 + A_3^T P_3 A_4 + \epsilon^2 A_3^T P_2^T A_2 = 0 \quad (5.54)$$

$$A_4^T P_3 A_4 - P_3 + Q_3 + \epsilon^2 \left(A_2^T P_1 A_2 + A_2^T P_2 A_4 + A_4^T P_2^T A_2 \right) = 0 \quad (5.55)$$

Define, like in Section 5.2, an $O(\epsilon^2)$ perturbation of (5.53)-(5.55) by

$$A_1^T \mathbf{P_1} A_1 - \mathbf{P_1} + Q_1 = 0 \quad (5.56)$$

$$A_1^T \mathbf{P_1} A_2 - \mathbf{P_2} + Q_2 + A_1^T \mathbf{P_2} A_4 + A_3^T \mathbf{P_3} A_4 = 0 \quad (5.57)$$

$$A_4^T \mathbf{P_3} A_4 - \mathbf{P_3} + Q_3 = 0 \quad (5.58)$$

Assume that the matrices A_1 and A_4 are asymptotically stable (Assumption 5.2). Then, the unique solutions of (5.56)-(5.58) exist.

Define the errors as

$$P_1 = \mathbf{P_1} + \epsilon^2 E_1, \; P_2 = \mathbf{P_2} + \epsilon^2 E_2, \; P_3 = \mathbf{P_3} + \epsilon^2 E_3 \qquad (5.59)$$

Subtracting (5.56)-(5.58) from (5.53)-(5.55), the following error equations are obtained

$$A_1^T E_1 A_1 - E_1 = -A_1^T P_2 A_3 - A_3^T P_2^T A_1 - A_3^T P_3 A_3 \qquad (5.60a)$$

$$A_4^T E_3 A_4 - E_3 = -A_2^T P_2 A_4 - A_4^T P_2^T A_2 - A_2^T P_3 A_2 \qquad (5.60b)$$

$$A_1^T E_2 A_4 - E_2 = -A_1^T E_1 A_2 - A_3^T E_3 A_4 - A_3^T P_2^T A_2 \qquad (5.60c)$$

The proposed parallel synchronous algorithm for the numerical solution of (5.60) is as follows.

Algorithm 5.4: (Shen et al., 1991).

$$A_1^T E_1^{(i+1)} A_1 - E_1^{(i+1)} = -A_1^T P_2^{(i)} A_3 - A_3^T P_2^{(i)^T} A_1 - A_3^T P_3^{(i)} A_3 \quad (5.61a)$$

$$A_4^T E_3^{(i+1)} A_4 - E_3^{(i+1)} = -A_2^T P_2^{(i)} A_4 - A_4^T P_2^{(i)^T} A_2 - A_2^T P_3^{(i)} A_2 \quad (5.61b)$$

$$A_1^T E_2^{(i+1)} A_4 - E_2^{(i+1)} = -A_1^T E_1^{(i+1)} A_2 - A_3^T E_3^{(i+1)} A_4 - A_3^T P_2^{(i)^T} A_2 \qquad (5.61c)$$

with starting points $E_1^{(0)} = E_2^{(0)} = E_3^{(0)} = 0$ and

$$P_j^{(i)} = \mathbf{P_j} + \epsilon^2 E_j^{(i)}, \quad j = 1, 2, 3; \; i = 1, 2, \ldots \qquad (5.62)$$

$$\triangle$$

Now we have the following theorem (Shen et al., 1991).

Theorem 5.4 *Under the stability Assumption 5.2, the algorithm (5.61)-(5.62) converges to the exact solutions for $E_j, j = 1, 2, 3$, with the rate of convergence of $O(\epsilon^2)$.*

$$\square$$

The proof of this theorem is similar to the proof of the corresponding algorithm for the continuous-time algebraic singularly perturbed Lyapunov equation studied in Section 5.2. It uses the bilinear transformation from (Power, 1967) to transform the discrete-time Lyapunov equation into the continuous one and then follows the ideas of Section 5.2.

Algorithm 5.4 is extended in (Shen and Gajic, 1990a) to the problem of solving the algebraic Riccati equation of weakly coupled systems in terms of the reduced-order subsystems.

5.4.2 Case Study: Discrete Catalytic Cracker

A fifth-order model of a catalytic cracker reactor (Kando et al., 1988), demonstrates the efficiency of the proposed method. The problem matrix A (after performing discretization with sampling period $T = 1$) is given by

$$A = \begin{bmatrix} 0.011771 & 0.046903 & 0.096679 & 0.071586 & -0.019178 \\ 0.014096 & 0.056411 & 0.115070 & 0.085194 & -0.022806 \\ 0.066395 & 0.252260 & 0.580880 & 0.430570 & -0.116280 \\ 0.027557 & 0.104940 & 0.240400 & 0.178190 & -0.048104 \\ 0.000564 & 0.002644 & 0.003479 & 0.002561 & -0.000656 \end{bmatrix}$$

The small weak coupling parameter ϵ is $\epsilon = 0.21$, and the matrix Q is chosen as identity.

The simulation results, obtained by using the software package L-A-S , are presented in Table 5.3. It can be seen from this table that the presented algorithm is numerically very efficient.

5.5 Summary

This chapter considers two special classes of systems, and solutions are presented for the associated algebraic Lyapunov matrix equations. These systems are the singularly perturbed and weakly coupled linear systems. It has been shown that the Lyapunov equations for these systems can be decomposed into three lower order Lyapunov equations. In order to obtain numerical solutions for such lower-order Lyapunov equations the recursive methods are presented. For weakly coupled systems these recursive methods can be performed in parallel because the reduced-order equations are completely decoupled. Even for singularly perturbed systems parallelism can be obtained to some degree. The most important feature of these recursive algorithms is that they yield the solution with an arbitrary degree of accuracy. If k represents the number of iterations performed then the algorithms for singularly perturbed systems give the accuracy of $O(\epsilon^k)$ while for weakly coupled systems the accuracy is $O(\epsilon^{2k})$.

i	$P_1^{(i)}$	$P_2^{(i)}$	$P_3^{(i)}$
0	1.0003 0.001354 1.005400	0.54689 0.40537 -0.10944 2.08640 1.54650 -0.41752	1.9302 0.68954 -0.18620 1.51110 -0.13802 1.03730
1	1.0139 0.052897 1.201800	0.66593 0.49359 -0.13322 2.54040 1.88290 -0.50820	2.2032 0.89183 -0.24071 1.66100 -0.17841 1.04820
2	1.0162 0.061844 1.236000	0.69091 0.51209 -0.13821 2.63570 1.95350 -0.52720	2.2601 0.93400 -0.25208 1.69230 -0.18683 1.05040
3	1.0167 0.063711 1.243100	0.69604 0.51590 -0.13923 2.65520 1.96800 -0.53113	2.2717 0.94260 -0.25439 1.69860 -0.18683 1.05040
4	1.0168 0.064092 1.244500	0.69710 0.51668 -0.13944 2.65930 1.97100 -0.53193	2.2741 0.94437 -0.25487 1.70000 -0.18891 1.05100
5	1.0168 0.064170 1.244800	0.69731 0.51684 -0.13948 2.66010 1.97160 -0.53210	2.2746 0.94473 -0.25497 1.70020 -0.18898 1.05100
6	1.0168 0.064186 1.244900	0.69736 0.51687 -0.13949 2.66010 1.97170 -0.53213	2.2747 0.94481 -0.25499 1.70030 -0.18899 1.05100
7	1.0168 0.064190 1.244900	0.69737 0.51688 -0.13950 2.66030 1.97180 -0.53214	2.2747 0.94482 -0.25499 1.70030 -0.18900 1.05100

$$P^{(7)} = P_{exact}$$

Table 5.3: Solution of the weakly coupled discrete algebraic Lyapunov equation

5.6 References

1. Bertsekas, D. and J. Tsitsiklis, *Parallel and Distributed Computation: Numerical Methods*, Prentice Hall, Englewood Cliffs, 1989.

2. Bertsekas, D. and J. Tsitsiklis, "Some aspects of parallel and distributed algorithms - a survey," *Automatica*, vol.27, 3-21, 1991.

3. Blankenship, G., "Singularly perturbed difference equations in optimal control problems," *IEEE Trans. Automatic Control*, vol.26, 911–917, 1981.

4. Bingulac, S. and H. Vanlandingham, *Algorithms for Computer-Aided Design of Multivariable Control Systems*, Marcel Dekker, New York, 1993.

5. Butuzov, V. and A. Vasileva, "Differential and difference equation systems with a small parameter for the case in which the unperturbed (singular) system is in the spectrum," *J. Differential Equations*, vol.6, 499-510, 1971.

6. Derbel, N., M. Kamoun, and M. Poloujadoff, "New approach to block-diagonalization of singularly perturbed systems by Taylor expansion," *IEEE Trans. Automatic Control*, vol.39, 1429–1431, 1994.

7. Gajic, Z., D. Petkovski, and N. Harkara, "The recursive algorithm for the optimal static output feedback control problem of linear singularly perturbed systems," *IEEE Trans. Automatic Control*, vol.34, 465-468, 1989.

8. Gajic, Z., D. Petkovski, and X. Shen, *Singularly Perturbed and Weakly Coupled Linear Control Systems: A Recursive Approach*, Lecture Notes in Control and Information Sciences, Springer Verlag, Berlin, 1990.

9. Gajic, Z. and X. Shen, *Parallel Algorithms for Optimal Control of Linear Large Scale Systems*, Springer Verlag, London, 1993.

10. Hoppensteadt, F. and W. Miranker, "Multitime methods for systems of difference equations," *Studies Appl. Math.*, vol.56, 273-289, 1977.

11. Kando, H., T. Iwazumi, and H. Ukai, "Singular perturbation modeling of large scale system with multitime scale property," *Int. J. Control*, vol.48, 2361–2387, 1988.

12. Khorasani, K. and M. Azimi-Sadjadi, "Feedback control of two time scale block implemented discrete time systems," *IEEE Trans. Automatic Control*, vol.32, 69–73, 1987.

13. Kokotovic, P., W. Perkins, J. Cruz Jr., and G. D'Ans, "ϵ-Coupling for near-optimum design of large scale linear systems," *Proc. IEE, Part D.*, vol.116, 889–892, 1969.

14. Kokotovic, P. and H. Khalil, *Singular Perturbations in Systems and Control*, IEEE Press, 1986.

15. Kokotovic, P., H. Khalil, and J. O'Reilly, *Singular Perturbation Methods in Control: Analysis and Design*, Academic Press, 1986.

16. Litkouhi, B., *Sampled-Data Control of Systems with Slow and Fast Models*, Ph. D. Dissertation, Michigan State University, 1983.

17. Litkouhi, B. and H. Khalil, "Infinite-time regulators for singularly perturbed difference equations," *Int. J. Control*, vol.39, 587–598, 1984.

18. Litkouhi, B. and H. Khalil, "Multirate and composite control of two-time-scale discrete systems," *IEEE Trans. Automatic Control*, vol.30, 645–651, 1985.

19. Luenberger, D., *Optimization With Vector Space Methods*, Wiley, New York, 1969.

20. Naidu, D. and Rao, *Singular Perturbation Analysis of Discrete Control Systems*, Lecture Notes in Mathematics, Springer Verlag, New York, 1985.

21. Mahmoud, M., Y. Chen, and M. Singh, "Discrete two-time-scale systems," *Int. J. Systems Science*, vol.17, 1187–1207, 1986.

22. Oloomi, H. and M. Sawan, "The observer-based controller design of discrete time singularly perturbed systems," *IEEE Trans. Automatic Control*, vol.32, 246–248, 1987.

23. Phillips, R., "Reduced order modeling and control of two time scale discrete control systems," *Int. J. Control*, vol.31, 761–780, 1980.

24. Power, H., "Equivalence of Lyapunov matrix equations for continuous and discrete systems," *Electronics Letters*, vol.3, 83, 1967.

25. Qureshi, M., X. Shen, and Z. Gajic, "Optimal output feedback control of linear singularly perturbed stochastic systems," *Int. J. Control*, vol.55, 361–371, 1992.

26. Shen, X., *Near-Optimum Reduced-Order Stochastic Control of Linear Discrete and Continuous Systems with Small Parameters*, Ph. D. Dissertation, Rutgers University, 1990.

27. Shen, X. and Z. Gajic, "Optimal reduced order solution of the weakly coupled discrete Riccati equation," *IEEE Trans. Automatic Control*, vol.35, 600–602, 1990a.

28. Shen, X. and Z. Gajic, "Approximate parallel controllers for discrete weakly coupled linear stochastic systems," *Optimal Control Appl. & Methods*, vol.11, 345–354, 1990b.

29. Shen, X., Z. Gajic, and D. Petkovski, "Parallel reduced-order algorithms for Lyapunov equations of large scale systems," *Proc. IMACS Symp.*, 697–702, Lille, France, 1991.

Chapter Six

Stability Robustness and Sensitivity of Lyapunov Equation

In this chapter we study two important problems related to the solution of the algebraic Lyapunov equation. The stability robustness of linear time invariant systems can be studied in terms of the solution of the algebraic Lyapunov equation, which was demonstrated in the papers by (Patel and Toda, 1980; Yedavalli, 1985a; Zhou and Khargonekar, 1987). In the second part of this chapter, we examine sensitivity of the solution of the algebraic Lyapunov equations in both continuous-time and discrete-time domains following results of (Hewer and Kenney, 1988; Gahinet et al., 1990).

6.1 Stability Robustness

Consider a time invariant asymptotically stable linear system

$$\dot{x}(t) = Ax(t), \qquad x(t_0) = x_0 \tag{6.1}$$

Assume that this system is perturbed (disturbed) by $f(x(t), t)$ with $f(0, t) = 0$, that is

$$\dot{x}(t) = Ax(t) + f(x(t), t), \qquad x(t_0) = x_0 \tag{6.2}$$

The fundamental stability robustness problem is to determine the magnitude of perturbation $f(x(t), t)$ such that the system (6.2) remains stable.

The answer to this problem is obtained in the work of (Patel and Toda, 1980) in the form of the following theorem.

Theorem 6.1: *The system (6.2) is stable if*

$$\frac{\|f(x(t,t))\|}{\|x(t)\|} \leq \mu = \frac{\lambda_{min}(Q)}{\lambda_{max}(P)}$$

where P satisfies the algebraic Lyapunov equation

$$A^T P + P A = -2Q$$

□

It has been shown in (Patel and Toda, 1980) that the bound given in Theorem 6.1 is maximal for $Q = I$, which implies the following very well-known lemma widely used in the control engineering literature.

Lemma 6.1: *The system (6.2) is stable if*

$$\frac{\|f(x(t,t))\|}{\|x(t)\|} \leq \mu = \frac{1}{\lambda_{max}(P)} \tag{6.3}$$

where P satisfies the algebraic Lyapunov equation

$$A^T P + P A = -2I \tag{6.4}$$

□

In a particularly important case of linear perturbations where

$$f(x(t), t) = E(t)x(t) \tag{6.5}$$

the following corollary of Theorem 6.1 is established in (Patel and Toda, 1980).

Corollary 6.1 *The linear system*

$$\dot{x}(t) = (A + E(t))x(t) \tag{6.6}$$

is stable if any of the following conditions is satisfied

$$a) \quad \|E(t)\| \leq \mu, \qquad b) \quad \|E(t)\|_F \leq \mu$$

$$\tag{6.7}$$

$$c) \quad |e_{ij}(t)| \leq \frac{\mu}{n} = \frac{1}{n\sigma_{max}(P)} = \frac{1}{n\lambda_{max}(P)}$$

where $\|.\|$ *denotes the spectral norm,* F *stands for Frobenius norm,* n *is the dimension of the matrix* A, *and* μ *is defined in (6.3).*

$\square$

Results of (Patel and Toda, 1980) do not exploit the structure of perturbations. This has been done in the work of (Yedavalli, 1985a), where the stability robustness bound estimate obtained in (6.7) part c) is improved. The obtained result for structured perturbations is stated in the following theorem.

Theorem 6.2: *The time invariant system (6.6) is stable if*

$$|e_{ij}| \leq \varepsilon < \frac{1}{\sigma_{max}[|P|U_n]_s} \tag{6.8}$$

where P *satisfies the algebraic Lyapunov equation (6.4),* U_n *is an* $n \times n$ *matrix with all elements equal to 1,* $|P|$ *indicates that all elements of matrix* P *are replaced by their absolute values,* $[.]_s$ *denotes the symmetric part of the corresponding matrix, and* σ_{max} *represents the maximal singular value.*

$\square$

It has been shown in (Yedavalli, 1985a) that

$$\sigma_{max}[|P|U_n]_s \leq n\sigma_{max}(P) \tag{6.9}$$

which analytically proves that the bound in (6.8) is better than the one from (6.7) part c).

Example 6.1: Consider an approximate model of order four of an one-dimensional diffusion process with the symmetric system matrix (Sage and White, 1977)

$$A = \begin{bmatrix} -1 & 1 & 0 & 0 \\ 1 & -2 & 1 & 0 \\ 0 & 1 & -2 & 1 \\ 0 & 0 & 1 & -1 \end{bmatrix}$$

The eigenvalues of the matrix A are given by $-0.1206, -1.0000,$ $-2.3473,$ $-3.5371,$ which indicates asymptotic stability of A. It is interesting to observe that due to the symmetry of A, the solution of the algebraic Lyapunov equation (6.4) has a very simple analytical form given by $P = -A^{-1}$. Using MATLAB we get

$$P = \begin{bmatrix} 4 & 3 & 2 & 2 \\ 3 & 3 & 2 & 1 \\ 2 & 2 & 2 & 1 \\ 1 & 1 & 1 & 1 \end{bmatrix}, \quad \lambda(P) = \begin{cases} 8.2909 \\ 1.0000 \\ 0.4260 \\ 0.2831 \end{cases}$$

Corollary 6.1 implies that the maximally allowed perturbations of the elements of the matrix A are

$$|e_{ij}| \leq \frac{1}{n\lambda_{max}(P)} = 0.0302, \quad i,j = 1,2,...,n$$

It is easy to check that the perturbed system matrix with $e_{ij} = 0.03$

$$A_p = A + 0.03 \begin{bmatrix} 1 & 1 & 1 & 1 \\ 1 & 1 & 1 & 1 \\ 1 & 1 & 1 & 1 \\ 1 & 1 & 1 & 1 \end{bmatrix} = \begin{bmatrix} -0.97 & 1.03 & 0.03 & 0.03 \\ 1.03 & -1.97 & 1.03 & 0.03 \\ 0.03 & 1.03 & -1.97 & 1.03 \\ 0.03 & 0.03 & 1.03 & -0.97 \end{bmatrix}$$

is asymptotically stable with the eigenvalues

$$\lambda(A_p) = \{-0.0122, -0.9911, -2.3451, -3.5317\}, \quad e_{ij} = 0.03$$

However, perturbing all the elements of the matrix A by $e_{ij} = 0.04$ causes instability. It can be concluded, in this particular example, that the perturbation bounds of Corollary 6.1 are pretty sharp. Interestingly enough, the results of Theorem 6.2 improves the obtained bounds even more. From (6.8) we get

$$|e_{ij}| \leq \frac{1}{\sigma_{max}[|P|U_n]_s} = 0.0333, \quad i,j = 1,2,...,n$$

By perturbing every element of the matrix A by $e_{ij} = 0.0332$ the asymptotic stability is preserved since the eigenvalues of the perturbed matrix are given by

$$\lambda(A_p) = \{-0.0005, -0.9902, -2.3448, -3.5316\}, \quad e_{ij} = 0.0332$$

It should be pointed out that the original system matrix is symmetric, which may be the reason why in this example the theoretical bounds almost coincide with the actual stability bounds.

$$\Delta$$

In the follow-up paper (Yedavalli, 1985b), it has been shown that the matrix U_n can be chosen as $U = \{\varepsilon_{ij}/\varepsilon\}$ assuming that the corresponding structural perturbation of matrix A is available. Stability robustness results obtained in (Yedavalli, 1985a, 1985b) were applied to the design of linear regulators (Yedavalli et al., 1985) and to stability analysis of interval matrices, (Yedavalli, 1986). In (Yedavalli and Liang, 1986) it has been indicated that the stability robustness bounds can be even more improved by using the state transformation and solving the corresponding algebraic Lyapunov equation in the new coordinates.

Generalization of the results of (Yedavalli, 1985a, 1985b) to an important class of structured perturbations appearing in the feedback control systems is given by (Zhou and Khargonekar, 1987). In their work it is assumed that the perturbation (uncertainty) matrix has the form

$$E = \sum_{i=1}^{m} k_i E_i \qquad (6.10)$$

where k_i are uncertain parameters. The main result of (Zhou and Khargonekar, 1987) is given in the next theorem.

Theorem 6.3: *The time invariant version of the system (6.6) is stable under structured perturbations defined in (6.10) if any of the following conditions is satisfied.*

$$\sum_{i=1}^{m} k_i^2 < \frac{1}{\sigma_{max}^2(P_e)} \qquad (6.11a)$$

$$\sum_{i=1}^{m} |k_i|\sigma_{max}(P_i) < 1 \qquad (6.11b)$$

$$|k_j| < \frac{1}{\sigma_{max}\left(\sum_{i=1}^{m} |P_i|\right)}, \qquad j = 1, 2, ..., m$$

where the matrix P satisfies (6.4) and P_i and P_e are given by

$$P_i = \frac{1}{2}\left(PE_i + E_i^T P\right) = [PE_i]_s, \quad i = 1, 2, ..., m$$
$$P_e = [P_1 \quad P_2 \quad \cdots \quad P_m] \qquad (6.12)$$

$$\square$$

Proof: Defining a Lyapunov function for the time invariant version of the system (6.6) as $V = x^T P x$, we get

$$\frac{dV}{dt} = 2x^T \left(\sum_{i=1}^{m} k_i P_i - I \right) x \tag{6.13}$$

which is negative definite for

$$\sigma_{max} \left(\sum_{i=1}^{m} k_i P_i \right) < 1 \tag{6.14}$$

Using notation introduced in (6.12), we have

$$\sum_{i=1}^{m} k_i P_i = P_e[k_1 I \quad k_2 I \quad \ldots \quad k_m I] \tag{6.15}$$

and by requirement that

$$\sigma_{max} \left(\sum_{i=1}^{m} k_i P_i \right) \leq \sigma_{max}(P_e) \left(\sum_{i=1}^{m} k_i^2 \right) < 1 \tag{6.16}$$

relation (6.11a) follows. In addition, the fact

$$\sigma_{max} \left(\sum_{i=1}^{m} k_i P_i \right) \leq \sum_{i=1}^{m} |k_i| \sigma_{max}(P_i) < 1 \tag{6.17}$$

verifies (6.11b). Formula (6.11c) is proved by observing that

$$\sigma_{max} \left(\sum_{i=1}^{m} k_i P_i \right) \leq \sigma_{max} \left(\sum_{i=1}^{m} |k_i P_i| \right) \leq \max_{j} |k_j| \sigma_{max} \left(\sum_{i=1}^{m} |P_i| \right) \tag{6.18}$$

∎

It can be shown that the bound obtained in Theorem 6.2 is a special case of (6.11c), (Zhou and Khargonekar, 1987).

Comment: Note that the proof of Theorem 6.3 in fact assures the asymptotic stability of the perturbed system and that the original system must be asymptotically stable in order to guarantee the existence of a unique positive definite solution for P from the algebraic Lyapunov equation (6.4).

Results of (Zhou and Khargonekar, 1987) are extended in (Bernstein and Haddad, 1989) to the simultaneous robust stability and performance analysis study of linear systems. Stability robustness for normal matrices $(A^T A = AA^T)$ is considered in (Jonckheere, 1984).

It is important to indicate that the stability robustness of linear systems can be also studied in the frequency domain. The link between the frequency domain analysis and the Lyapunov equation is presented in (Mori, 1989).

Stability robustness of systems with time delays in terms of the algebraic Lyapunov equation is considered in (Cheres et al., 1989; Xu and Liu, 1994; Xu, 1994).

Remark: At the end of this section, we would like to point out that in a very recent paper (Chen and Han, 1994), the stability robustness result due to unstructural perturbations known from (Patel and Toda, 1980) has been improved. The main result of (Chen and Han, 1994), dual to the result of Theorem 6.1, states that the system (6.2) is stable if

$$\frac{\|f(x(t,t))\|}{\|x(t)\|} \leq \mu_c = \frac{\sigma_{min}(Q^{1/2})}{\sigma_{max}(Q^{-1/2}P)} \tag{6.19}$$

where P satisfies the algebraic Lyapunov equation

$$A^T P + PA = -2Q \tag{6.20}$$

The result given in (6.19)-(6.20) is derived using the same logic as one of (Patel and Toda, 1980). Taking a quadratic form for the Lyapunov function, that is

$$V(x) = x^T P x \tag{6.21}$$

we get

$$\dot{V}(x) = x^T (A^T P + PA) x + 2f^T P x \leq 0 \tag{6.22}$$

which is satisfied if for some positive definite matrix Q we have

$$f^T P x \leq x^T Q x \tag{6.23}$$

New observation made by (Chen and Han, 1994) over the one of (Patel and Toda, 1980) is to rewrite (6.23) as

$$\left(Q^{-1/2}Pf\right)^T Q^{1/2}x \le \left(Q^{1/2}x\right)^T \left(Q^{1/2}x\right) \tag{6.24}$$

from which it follows

$$\left\|Q^{-1/2}Pf\right\| \le \left\|Q^{1/2}x\right\| \tag{6.25}$$

Since

$$\left\|Q^{-1/2}Pf\right\| \le \sigma_{max}\left(Q^{-1/2}P\right)\|f\| \tag{6.26}$$

and

$$\left\|Q^{1/2}x\right\| \ge \sigma_{min}\left(Q^{1/2}\right)\|x\| \tag{6.27}$$

the sufficient condition for stability, given by (6.19)-(6.20), is easily obtained from (6.25)-(6.27).

It is also shown analytically in (Chen and Han, 1994) that their result is less conservative than the one of (Patel and Toda, 1980). In addition, an algorithm has been constructed for finding the improved stability robustness measure μ_c. Two examples done in (Chen and Han, 1994) show that the stability robustness bound is improved by at least 18% over the corresponding result of (Patel and Toda, 1980).

6.2 Sensitivity

Sensitivity of the Lyapunov equation is studied in several papers related to the numerical solution of the Lyapunov equation (Golub et al., 1979; Laub, 1979; Hammarling, 1982). It has been also independently studied in the work of (Jonckheere, 1984; Hewer and Kenney, 1988) for the continuous-time, and by (Gahinet et al., 1990) for the discrete-time domain.

6.2.1 Sensitivity of the Continuous Algebraic Lyapunov Equation

In this section we present the main results of (Hewer and Kenney, 1988) and indicate the related results obtained by other researchers.

Consider the continuous-time algebraic Lyapunov equation

$$A^T P + PA + Q = 0 \tag{6.28}$$

The perturbed form of this equation is given by

$$(A + \Delta A)^T (P + \Delta P) + (P + \Delta P)(A + \Delta A) + (Q + \Delta Q) = 0 \tag{6.29}$$

Assuming that the matrix A is asymptotically stable, the main result of (Hewer and Kenney, 1988) is given in the following theorem.

Theorem 6.4: *The sensitivity bound of the solution of the continuous-time algebraic Lyapunov equation (6.28) is given by*

$$\frac{\|\Delta P\|}{\|P + \Delta P\|} \le 2\|A + \Delta A\| \|H\| \left\{ \frac{\|\Delta A\|}{\|A + \Delta A\|} + \frac{\|\Delta Q\|}{\|Q + \Delta Q\|} \right\} \tag{6.30}$$

where the matrix H satisfies the following algebraic Lyapunov equation

$$A^T H + HA + I = 0 \tag{6.31}$$

$\square$

Note that the norm used is the spectral norm and that the above theorem is valid if the perturbations are such that none of denominators in (6.30) is equal to zero. The complete proof of this theorem is given below.

Proof: From equation (6.29) we have

$$A^T \Delta P + \Delta PA = \\ -[\Delta Q + \Delta A(P + \Delta P) + (P + \Delta P)\Delta A] \tag{6.32}$$

Solution of this equation is given by

$$\Delta P = \int_0^\infty e^{A^T t} [\Delta Q + \Delta A(P + \Delta P) + (P + \Delta P)\Delta A] e^{At} dt \tag{6.33}$$

Multiplying the above equation by the unit left and right singular vectors, it follows that

$$u^* \Delta P v = \|\Delta P\| \\ = \int_0^\infty u^* e^{A^T t} [\Delta Q + \Delta A(P + \Delta P) + (P + \Delta P)\Delta A] e^{At} v \, dt \tag{6.34}$$

where * denotes the complex conjugate transpose. The last equation implies

$$\|\Delta P\| \leq \{\|\Delta Q\| + 2\|\Delta A\|\|P + \Delta P\|\} \int_0^\infty \|e^{At}u\|\|e^{At}v\| dt \quad (6.35)$$

Note that the solution of the algebraic Lyapunov equation (6.31) is given by

$$H = \int_0^\infty e^{A^T t} e^{At} dt \quad (6.36)$$

so that

$$\int_0^\infty \|e^{At}u\|^2 dt = u^* \int_0^\infty e^{A^T t} e^{At} dt\, u = u^* H u \leq \|H\| \quad (6.37)$$

Similarly, by the same arguments

$$\int_0^\infty \|e^{At}v\|^2 dt = v^* H u \leq \|H\| \quad (6.38)$$

Using the Cauchy-Schwarz inequality we get

$$\int_0^\infty \|e^{At}u\|\|e^{At}v\| dt \leq \left\{\int_0^\infty \|e^{At}u\|^2 dt\right\}^{\frac{1}{2}} \left\{\int_0^\infty \|e^{At}v\|^2 dt\right\}^{\frac{1}{2}} \leq \|H\|$$
$$(6.39)$$

so that inequality (6.35) becomes

$$\|\Delta P\| \leq \{\|\Delta Q\| + 2\|\Delta A\|\|P + \Delta P\|\}\|H\| \quad (6.40)$$

The original perturbed equation (6.29) implies

$$\|Q + \Delta Q\| \leq 2\|A + \Delta A\|\|P + \Delta P\| \quad (6.41)$$

Combining the last two inequalities we get (6.30), the statement of the Theorem 6.4, and thus compete the required proof.

∎

The sharpness of the sensitivity bound stated in Theorem 6.4 is comparable to the corresponding sensitivity bounds obtained by using the standard perturbation analysis for linear systems of algebraic equations (Golub et al., 1979; Hammarling, 1982).

The sensitivity of the algebraic Lyapunov equation having normal system matrix is studied in (Laub, 1979; Hammarling, 1982; Jonckheere, 1984).

6.2.2 Sensitivity of the Discrete Algebraic Lyapunov Equation

Sensitivity of the solution of the discrete-time algebraic Lyapunov equation due to parameter variations in the problem coefficients is considered in (Gahinet et al., 1990). For the given discrete algebraic Lyapunov equation

$$P = A^T P A + Q \tag{6.42}$$

the perturbed equation is given by

$$(P + \Delta P) = (A + \Delta A)^T (P + \Delta P)(A + \Delta A) + (Q + \Delta Q) \tag{6.43}$$

Assuming that the system matrix A is *asymptotically stable*, the following theorem, dual to Theorem 6.4, is obtained for the sensitivity of the discrete algebraic Lyapunov equation (Gahinet et al., 1990).

Theorem 6.5: *The sensitivity bound of the solution of the discrete-time algebraic Lyapunov equation (6.42) is given by*

$$\frac{\|\Delta P\|}{\|P + \Delta P\|} \le \|H\| \left\{ K_1 \frac{\|\Delta A\|}{\|A + \Delta A\|} + K_2 \frac{\|\Delta Q\|}{\|Q + \Delta Q\|} \right\} \tag{6.44}$$

where the matrix H satisfies the following discrete algebraic Lyapunov equation

$$H = A^T H A + I \tag{6.45}$$

and K_1 and K_2 are given by

$$K_1 = 2(\|F\| + \|\Delta F\|)^2, \quad K_2 \le 1 + \|F + \Delta F\|^2 \tag{6.46}$$

$\square$

It has been also shown in the same paper that

$$\|H\| = \|\mathcal{L}^{-1}\| \qquad (6.47)$$

where $\mathcal{L}$ is the discrete-time Lyapunov operator defined by

$$\mathcal{L}(P) = P - A^T P A \qquad (6.48)$$

The sensitivity results for both continuous-time and discrete-time algebraic Lyapunov equations obtained by (Hewer and Kenney, 1988) and (Gahinet et al., 1990) are very comprehensive, and the reader is referred to those papers for more details and related results.

6.3 References

1. Bernstein, D. and W. Haddad, "Robust stability and performance analysis for linear dynamic systems," *IEEE Trans. Automatic Control*, vol.34, 751–758, 1989.

2. Gahinet, P., A. Laub, C. Kenney, and G. Hewer, "Sensitivity of the stable discrete-time Lyapunov equation," *IEEE Trans. Automatic Control*, vol.35, 1209–1217, 1990.

3. Chen, H. and K. Han, "Improved quantitative measures of robustness for multivariable systems," *IEEE Trans. Automatic Control*, vol.39, 807–810, 1994.

4. Cheres, E., Z. Palmor, and S. Gutman, "Quantitative measures of robustness for systems including delayed perturbations," *IEEE Trans. Automatic Control*, vol.34, 1203–1204, 1989.

5. Golub, G., S. Nash, and C. Loan, "A Hessenberg-Schur method for the problem $AX + XB = C$," *IEEE Trans. Automatic Control*, vol.24, 909-913, 1979.

6. Hammarling, S., "Numerical solution of the stable, non-negative definite Lyapunov equation," *IMA J. Numerical Analysis*, vol.2, 303-323, 1982.

7. Hewer, G. and C. Kenney, "The sensitivity of the Lyapunov equation," *SIAM J. Control and Optimization*, vol.26, 321-344, 1988.

8. Jonckheere, E., "New bounds on the sensitivity of the solution of the Lyapunov equation," *Linear Algebra and Its Appl.*, vol.60, 57–64, 1984.

9. Laub, A., "A Schur method for solving algebraic Riccati equations," *IEEE Trans. Automatic Control*, vol.24, 913–921, 1979.

10. Mori, T., "Estimates for a measure of stability robustness via a Lyapunov matrix equation," *Int. J. Control*, vol.50, 435–438, 1989.

11. Patel, R. and M. Toda, "Quantitative measures of robustness for multi-variable systems," *Proc. Joint American Control Conf.*, paper TP8–A, San Francisco, 1980.

12. Sage, A. and C. White, *Optimum Systems Control*, Prentice Hall, Englewood Cliffs, 1977.

13. Yedavalli, R., "Improved measures of stability robustness for linear state space models," *IEEE Trans. Automatic Control*, vol.42, 577–579, 1985a.

14. Yedavalli, R., "Perturbation bounds for robust stability in linear state space models," *Int. J. Control*, vol.42, 1507–1517, 1985b.

15. Yedavalli, R., S. Banda, and D. Ridgely, "Time-domain stability robustness measure for linear regulators," *AIAA J. Guidance*, vol.8, 520–524, 1985.

16. Yedavalli, R. and Z. Liang, "Reduced conservatism in stability robustness bounds by state transformation," *IEEE Trans. Automatic Control*, vol.31, 863–866, 1986.

17. Yedavalli, R., "Stability analysis of interval matrices: another sufficient condition," *Int. J. Control*, vol.43, 767-772, 1986.

18. Xu, B. and Y. Liu, "An improved Razumikhin-type theorem and its applications," *IEEE Trans. Automatic Control*, vol.39, 839–841, 1994.

19. Xu, B., "Correction to "An improved Razumikhin-type theorem and its applications," *IEEE Trans. Automatic Control*, vol.39, 2368, 1994.

20. Zhou, K. and P. Khargonekar, "Stability robustness for linear state space models with structured uncertainty," *IEEE Trans. Automatic Control*, vol.32, 621-623, 1987.

Chapter Seven

Iterative Methods and Parallel Algorithms

Iterative methods and parallel algorithms for solving the algebraic Lyapunov equation are very useful for large scale systems and systems with sparse matrices.

Within the context of iterative methods, we consider the classic work of (R. Smith, 1968), and modern techniques based on the so-called alternating-direction-implicit (ADI) iteration scheme (Wachspress, 1988; Lu and Wachspress, 1991) and the block successive overrelaxation method (Starke, 1991a, 1992). The ADI method is particularly very efficient when the sparse matrices have small bandwidth and real spectra. It does show some difficulties with complex spectra, but the theory is still developing and the obtained results are very promising.

Presented parallel algorithm for solving continuous-time algebraic Lyapunov equation with block diagonally dominant matrix A is based on the work of (Hodel and Poolla, 1992). The parallel algorithms for solving both continuous-time and discrete-time coupled algebraic Lyapunov equations arising in the optimal control of jump linear systems are derived according to the results reported in (Borno, 1995a; Borno and Gajic, 1995). In both cases the coupled algebraic Lyapunov equations are solved iteratively in terms of the reduced-order decoupled algebraic Lyapunov equations.

7.1 Smith's Algorithm

An interesting and important iterative algorithm for solving the continuous-time algebraic Lyapunov equation has been developed by (R. Smith, 1968). For general large scale systems it is comparable to the modern ADI method.

The algebraic Lyapunov equation under consideration given by

$$A^T P + P A + Q = 0 \qquad (7.1)$$

can be easily rewritten in the form

$$(qI - A^T) P (qI - A) - (qI + A^T) P (qI + A) = 2qQ \qquad (7.2)$$

where q is any nonzero real scalar parameter. Premultiplying by $(qI - A^T)^{-1}$ and postmultiplying by $(qI - A)^{-1}$, we get (R. Smith, 1968)

$$P - V P V^T = W \qquad (7.3)$$

where

$$\begin{aligned} V &= (qI - A^T)^{-1} (qI + A^T) \\ W &= 2q(qI - A^T)^{-1} Q (qI - A)^{-1} \end{aligned} \qquad (7.4)$$

Note that (7.3)-(7.4) comprise the bilinear transformation (see Section 3.1.1) and that the obtained results can be used for solving the discrete-time algebraic Lyapunov equation as well. Assuming that the matrix A is asymptotically stable and that the parameter q is positive, the spectral radius of V is less than 1 (R. Smith, 1968) so that the following sequence is convergent

$$P^{(k+1)} = W + V P^{(k)} V^T \qquad (7.5)$$

This sequence converges to the unique solution of (7.1) and generates an infinite series of the form

$$P = \sum_{k=1}^{\infty} V^{k-1} W \left(V^{k-1} \right)^T \qquad (7.6)$$

The presented scheme (7.5) has linear convergence. It can be modified such that one gets the quadratic convergence as proposed by

(R. Smith, 1968). That algorithm requires squaring of the matrix V (doubling idea, nicely explained in (Anderson, 1978)) in each iteration step and is given below.

Algorithm 7.1:

1. For some scalar positive constant q construct the matrices (7.4) and take the initial condition as $P^{(0)} = W$.
2. For $k = 0, 1, 2, \ldots$ perform iterations

$$P^{(k+1)} = P^{(k)} + V^{2^k} P^{(k)} \left(V^{2^k} \right)^T \qquad (7.7)$$

$$\triangle$$

The number of iterations required for desired accuracy depends on the parameter q. Typical value of q for experiments performed by (P. Smith, 1971) was 0.1. According to (P. Smith, 1971) this algorithm requires $2.5n^2$ words of memory and $2.5n^3(k+1)$ multiplications and was successfully used to solve the algebraic Lyapunov equation of order $n = 146$. It should be pointed out that using the methodology from the next section an optimal value for the parameter q that assures rapid convergence can be obtained.

Example 7.1: Consider the fifth-order algebraic Lyapunov equation from Example 2.7. Here, we present the results for the Smith algorithm and examine the relation of the number of iterations on the parameter q such that the accuracy of $O(10^{-14})$ is achieved. Obtained results are given in Table 7.1.

q	0.1	1	5	10	20	30	50	100	1000
k	14	10	8	7	6	7	8	9	12

Table 7.1: Number of iterations as a function of parameter q

It is interesting to observe from Table 7.1 that, for this particular example, the algorithm is pretty much insensitive to the values of parameter q.

Note that the similar iterative algorithm for solving (7.1) with $q = 1$ and $Q = I$ was proposed by (Barnett and Storey, 1967, 1968). Also, the sequence (7.7) is obtained in (Davison and Man, 1968) while studying the problem of numerical solution for the continuous-time algebraic Lyapunov equation (see Section 2.3, Algorithm 2.4).

The Smith method in fact represents an ADI iterative scheme with a single parameter. It is very convenient for solving large order Lyapunov equations and is comparable to the modern ADI iterative methods (Wachspress, 1988; Lu and Wachspress, 1991; Starke, 1992) to be presented in the next section.

7.2 ADI Iterative Method

The simplest iterative methods for solving large algebraic Lyapunov equations could have been constructed by converting the Lyapunov equation (7.1) into a system of linear algebraic equations as presented in Section 2.1.1, that is

$$\mathcal{A}\mathbf{p} = -\mathbf{q}, \qquad \mathcal{A} = A^T \otimes I + I \otimes A^T$$

$$\mathbf{p} = \begin{bmatrix} p_1 \\ p_2 \\ \vdots \\ p_n \end{bmatrix}, \qquad \mathbf{q} = \begin{bmatrix} q_1 \\ q_2 \\ \vdots \\ q_n \end{bmatrix} \tag{7.8}$$

where $p_i, q_i, i = 1, 2, ..., n$, are row vectors of matrices P and Q, and then by using any of the standard methods (Young, 1971) for iterative solution of the obtained large system of linear algebraic equations (7.8), such as Jacobi, Gauss-Seidel, and successive overrelaxation methods. Among several known iterative methods two of them appear to be very promising for solving the large algebraic Lyapunov equation: alternating-direction-implicit (ADI) iterations and block successive overrelaxation (SOR) iterations.

Alternating-direction-implicit iteration method represents a generalization of the block Jacobi method. ADI iterative method is used for solving large scale Lyapunov equations when the system matrix A is either asymptotically stable or N-stable (real parts of the eigenvalues of the matrix A are strictly positive). Recall that the Bartels-Stewart algorithm presented in Section 2.3.1 is applicable under the assumption that the matrix A is nonsingular and is thus more general. However, in the case when the matrix A has small bandwidth and real (asymptotically stable or N-stable) spectrum the ADI algorithm is superior over the Bartels-Stewart algorithm (Wachspress, 1992). The complete results for ADI methods are obtained for real spectra. The theory of ADI algorithms with complex spectra is still developing and some valuable results have been already obtained (Wachspress, 1990; Ellner and Wachspress, 1991; Lu and Wachspress, 1991; Starke, 1991b).

An iterative method for solving the continuous-time algebraic Lyapunov equation based on ADI theory has been presented in (Wachspress, 1988; Lu and Wachspress, 1991).

Algorithm 7.2:

$$\left(A^T + p_j I\right) X^{(j-1/2)} = -Q - X^{(j-1)}(A - p_j I), \qquad X^{(0)} = 0$$
$$\left(A^T + p_j I\right) X^{(j)} = -Q - X^{(j-1/2)^T}(A - p_j I), \qquad j = 1, 2, ..., N$$
$$\tag{7.9}$$
$$\Delta$$

Note that the matrices $X^{(j)}$ are symmetric despite the fact that the matrices $X^{(j-1/2)}$ may be nonsymmetric.

Implementation of this algorithm for the case when matrix A is sparse can be found in (Wachspress, 1988).

The required computations for Algorithm 7.2 are $\frac{7}{2}n^3$ flops per iteration. In order to reduce computations in (Lu and Wachspress, 1991) the reduction to the tridiagonal form by the Gaussian elimination is performed first and then the obtained tridiagonal system is solved by the ADI technique. In the case when the system matrix A is tridiagonal, the ADI method requires only $12n^2$ flops per iteration. However, the reduction to the tridiagonal form and the solution recovery require additional $\frac{19}{3}n^3$ flops. The optimal values for p_j are obtained in terms of the spectrum parameters by using the theory of elliptic functions (Wachspress, 1988; Lu and Wachspress, 1991). Thus, an efficient method for estimating the spectrum of the system matrix A is required.

The theory of ADI method for solving large sparse algebraic Lyapunov equations with complex spectra is still developing. For the case of highly oscillatory and slightly damped system modes (eigenvalues of matrix A) the ADI method requires a lot of iterations (Hodel, 1992).

During the tridiagonalization procedure the breakdowns may happen so that the procedure is the problem dependent and in the case when only a few breakdowns occur the ADI technique is competitive to the Bartels-Stewart algorithm. However, the final word about the power of ADI methods is not said yet and they remain very promising tools for solving large sparse Lyapunov equations.

7.3 SOR Iterative Method

In addition to ADI methods, the successive overrelaxation method (SOR) is also very well suited for solving large algebraic Lyapunov equations as demonstrated in (Starke, 1991a). The classical successive overrelaxation method applied to (7.8) has the form (Young, 1971)

$$\mathbf{p}^{(k+1)} = \left\{ (\mathcal{D} - \omega\mathcal{L})^{-1}[(1 - \omega)\mathcal{D} + \omega\mathcal{U}] \right\} \mathbf{p}^{(k)} - \omega(\mathcal{D} - \omega\mathcal{L})^{-1}\mathbf{q} \tag{7.10}$$

where the system matrix $\mathcal{A}$ is decomposed as $\mathcal{A} = \mathcal{D} - \mathcal{L} - \mathcal{U}$ with $\mathcal{D}$ being diagonal, $\mathcal{L}$ (strictly) lower triangular, and $\mathcal{U}$ (strictly) upper triangular matrices. The parameter ω is chosen such that the fastest convergence is achieved. It is very well known that for the classic successive overrelaxation method the parameter ω belongs to $\omega \in (0, 2)$.

It has been shown in (Starke, 1991a) that the so-called block SOR method (Varga, 1962, page 78) is very efficient for solving the algebraic Lyapunov equation. For the algebraic filter type Lyapunov equation, which represents the transpose of equation (7.1), that is

$$AK + KA^T = Q \tag{7.11}$$

Starke has derived the following block SOR (Starke, 1991a, 1992) by using row-wise numbering.

Algorithm 7.3:

$$(D - \omega L)K^{(j)} + K^{(j)}A^T = \\ [(1 - \omega)D + \omega U]K^{(j-1)} + (1 - \omega)K^{(j-1)}A^T - \omega Q \tag{7.12}$$

with $A = D - L - U$, and D diagonal, L (strictly) lower triangular, and U (strictly) upper triangular parts of A.

Δ

A procedure for finding the optimal value for the parameter ω in Algorithm 7.3 is considered in (Starke, 1991).

A dual version of Algorithm 7.3 can be obtained by using column-wise numbering, that is

$$AK^{(j)} + K^{(j)}(D - \omega L^T) = \\ (1 - \omega)AK^{(j-1)} + K^{(j-1)}[(1 - \omega)D + \omega U^T] - \omega Q \tag{7.13}$$

The convergence features of the SOR method are improved in (Starke, 1992) by combining it with the ADI method and forming the so-called alternating direction SOR (ADSOR) method. This has been achieved by combining algorithms (7.12) and (7.13) in the spirit of alternating direction implicit iterations. The new ADSOR algorithm has the form.

Algorithm 7.4:

$$(D - \omega L)K^{(j-1/2)} + K^{(j-1/2)}A^T = \\ [(1 - \omega)D + \omega U]K^{(j-1)} + (1 - \omega)K^{(j-1)}A^T - \omega Q$$

$$\tag{7.14}$$

$$AK^{(j)} + K^{(j)}(D - \omega L^T) = \\ (1 - \omega)AK^{(j-1/2)} + K^{(j-1/2)}[(1 - \omega)D + \omega U^T] - \omega Q$$

$$\Delta$$

Computational experiments for Algorithms 7.3 and 7.4 have been performed in (Starke, 1992). It has been demonstrated that ADSOR converges even in a case for which SOR diverges for every $\omega \in (0, 2)$. However, the analysis for ADSOR is not complete yet and this algorithm will be the subject of future research (Starke, 1992).

7.4 Parallel Algorithms

Parallel version of the Bartels Stewart algorithm (see Section 2.3.1) has been discussed in (O'Leary and Stewart, 1985) for the case when the system matrix A is upper tridiagonal with the real spectrum. In (Hodel and Poolla, 1992) a parallel implementation of the Hammarling algorithm (see Section 2.3.1) for large and dense matrix A is considered.

In this section we present a parallel algorithm, the so-called full-rank perturbed iteration (FRPI) for solving the algebraic Lyapunov equation, which has been recently obtained by (Hodel and Poolla, 1992). This algorithm is applicable for large banded diagonally dominant system matrices satisfying $A + A^T < 0$. Such matrices appear in the analysis and design of large flexible structures (Balas, 1982).

For the algebraic Lyapunov "filter type" equation (7.11) a linear operator can be defined as

$$\mathbf{L}(K) = AK + KA^T \tag{7.15}$$

so that the solution of the equation (7.11) is given by

$$K = -\mathbf{L}^{-1}(Q) \tag{7.16}$$

The leading idea in (Hodel and Poolla, 1992) is to construct a perturbed block-diagonal algebraic Lyapunov equation, which has the same solution as the one in (7.11), that is

$$A_0 K + K A_0^T + Q + \Delta Q = 0 \tag{7.17}$$

It is assumed that the matrix A has N dominant diagonal blocks so that the matrix A_0 is selected as $A_0 = diag\{A_{11}, A_{22}, ..., A_{NN}\}$. In the operator form the solution of the last equation can be recorded as

$$K = -\mathbf{L}_0^{-1}(Q + \Delta Q) \tag{7.18}$$

Since the matrix A is a general banded matrix it can be represented as

$$A = A_0 + \Delta A \tag{7.19}$$

It is easy to observe from (7.11) and (7.17) that

$$(\Delta A)K + K(\Delta A)^T - \Delta Q = 0 \Rightarrow \Delta \mathbf{L}(K) = \Delta Q \tag{7.20}$$

which implies

$$K = \Delta \mathbf{L}^{-1}(\Delta Q) \tag{7.21}$$

The perturbation matrix ΔQ causes that the algebraic equations (7.11) and (7.17) have the same solution. This matrix can be determined from (7.18) and (7.21) as follows

$$\begin{aligned} \Delta Q = \Delta \mathbf{L}(K) &= -\Delta \mathbf{L}\big(\mathbf{L}_0^{-1}(Q + \Delta Q)\big) \\ &= -\Delta \mathbf{L}\big(\mathbf{L}_0^{-1}(Q)\big) - \Delta \mathbf{L}\big(\mathbf{L}_0^{-1}(\Delta Q)\big) \end{aligned} \tag{7.22}$$

It has been proposed to solve this equation by fixed-point iterations, that is

$$\begin{aligned} \Delta Q^{(k+1)} &= -\Delta \mathbf{L}\Big(\mathbf{L}_0^{-1}\big(\Delta Q^{(k)}\big)\Big) - \Delta \mathbf{L}\big(\mathbf{L}_0^{-1}(Q)\big) \\ &= -\Delta \mathbf{L}\Big(K^{(k)}\Big) - q \end{aligned} \tag{7.23}$$

where q is a constant term and $K^{(k)}$ satisfies

$$A_0 K^{(k)} + K^{(k)} A_0^T = \Delta Q^{(k)} \tag{7.24}$$

Note that due to the block-diagonal structure of the matrix A_0, equation (7.24) can be solved in parallel in terms of reduced-order problems by using the Bartels-Stewart algorithm leading to

$$A_{ii} K_{ij}^{(k)} + K_{ij}^{(k)} A_{jj}^T = \Delta Q_{ij}^{(k)} \tag{7.25}$$

It has been proved that the proposed fixed-point iterations (7.23) converge under the *weak coupling assumption*, that is, "the "coupling" blocks ΔA are sufficiently small," (Hodel and Poolla, 1992).

The FRPI algorithm of (Hodel and Poolla, 1992) can be now summarized as follows.

Algorithm 7.5:

1. Set $\Delta Q^{(0)} = 0, k = 0$.
2. Select a block-diagonal matrix A_0 and evaluate the constant term q in (7.23) by solving in parallel, like in (7.25), equation $A_0 K_0 + K_0 A_0^T + Q = 0$ and by using $q = (\Delta A)K_0 + K_0(\Delta A)^T$.
3. Solve in parallel, like in (7.25), equation $A_0 K^{(k)} + K^{(k)} A_0^T + \Delta Q^{(k)} = 0$ and evaluate in parallel $\Delta \mathbf{L}\big(K^{(k)}\big)$.
4. Evaluate $\Delta Q^{(k+1)}$ from (7.23) now as $\Delta Q^{(k+1)} = -\Big[(\Delta A)K^{(k)} + K^{(k)}(\Delta A)^T\Big] - q$.
5. Stop if $\big\| \Delta Q^{(k+1)} - \Delta Q^{(k)}\big\| \leq \varepsilon$, where ε indicates the required accuracy. Otherwise take $k = k + 1$ and repeat steps 3–5.

$$\Delta$$

Note that the solution generated by the sequence of the algebraic Lyapunov equations in Step 3, that is, $K^{(k)}$ converges to the desired solution of the algebraic Lyapunov equation (7.11).

7.5 Parallel Algorithm for Coupled Lyapunov Equations

Systems of coupled Lyapunov equations play a very important role in the study of linear jump parameter systems with Markovian transitions (Mariton, 1990). They are encountered in stability analysis, optimal control and filtering problems, and in computational algorithms such as the Newton method for solving algebraic coupled Riccati equations of jump linear systems, (Wonham, 1971). Exact solution of coupled differential Lyapunov equations of jump linear systems has been obtained in (Jodar and Mariton, 1987). The solution is based on the Kronecker product representation, hence, it suffers from high dimensionality and, consequently, is not suitable for practical applications. In this section we present two recursive reduced-order parallel algorithms for solving the corresponding continuous-time and discrete-time coupled *algebraic* Lyapunov equations, by following the work of (Borno, 1995a, 1995b; Borno and Gajic, 1995).

7.5.1 Continuous Coupled Algebraic Lyapunov Equations[1]

The coupled continuous algebraic Lyapunov equations of jump linear systems are given by

$$A_j^T K_j + K_j A_j + Q_j + \sum_{l=1}^{N} p_{jl} K_l = 0, \quad j = 1, 2, ..., N \quad (7.26)$$

The constant matrices A_j are of dimensions $n \times n$. Matrices $Q_j = Q_j^T \in \Re^{n \times n}$ are positive semi-definite so that the solution matrices K_j

---

[1]   Reprinted with minor changes from *Automatica*, vol.31, I. Borno, "Parallel computation of the solutions of coupled algebraic Lyapunov equations," in press, 1995, with kind permission from Elsevier Science Ltd, The Boulevard, Langford Lane, Kidlington OX5 1GB, UK.

are symmetric. Parameters p_{ij} satisfy $p_{ij} > 0, i \neq j, p_{jj} = -\sum_{i \neq j} p_{ij}$, (Kumar and Varaiya, 1986). More about jump linear systems can be found in Section 8.2.

Assume that equations (7.26) have the unique positive semi-definite solutions $K_j, j = 1, 2, ..., N$, which is the case when the matrices $A_j, j = 1, 2, ..., N$, are stochastically stable (Ji and Chizeck, 1990). The algorithm to be proposed is valid under the following assumption.

Assumption 7.1 The matrices $A_j, j = 1, 2, ..., N$, are asymptotically stable.

Δ

The following algorithm for solving (7.26) is proposed by (Borno, 1995a, 1995b).

Algorithm 7.6:

$$\mathbf{A_j^T} K_j^{(i+1)} + K_j^{(i+1)} \mathbf{A_j} + Q_j + \sum_{l=1, l \neq j}^{N} p_{lj} K_l^{(i)} = 0 \qquad (7.27)$$

with $\mathbf{A_j} = A_j + \frac{1}{2} p_{jj} I$, and initial conditions obtained from

$$A_j^T K_j^{(0)} + K_j^{(0)} A_j + Q_j = 0, \quad j = 1, 2, ..., N \qquad (7.28)$$

Δ

Matrices $\mathbf{A_j}$ are stable by Assumption 7.1 and the fact that $p_{jj} < 0$. It can be noticed that the proposed algorithm is very efficient since it operates on the decoupled Lyapunov equations. The following theorem indicates the main features of Algorithm 7.6.

Theorem 7.1 *Assume that the unique positive semi-definite solutions of (7.26) exist and that Assumption 7.1 is satisfied. Then, Algorithm 7.6 converges to the unique positive semi-definite solution of the coupled algebraic Lyapunov equations (7.26).*

$\square$

Proof: Since $\mathbf{A_j}$ are stable matrices and Q_j are positive semi-definite we have that $K_j \geq 0, j = 1, 2, ..., N$. From (7.26) and (7.28) it follows that $K_j \geq K_j^{(0)}, \forall j$. Thus, the algorithm produces

$$K_j \geq ... \geq K_j^{(i)} \geq K_j^{(i-1)} \geq ... \geq K_j^{(0)} \geq 0, \quad j = 1, 2, ..., N \quad (7.29)$$

It can be seen from the last relations that the sequence of solutions of (7.27) is monotonically nondecreasing with K_j representing the upper bound. Thus, this sequence is convergent by the theorem of monotonic convergence of positive operators (Kantorovich and Akilov, 1964). Assume that the limit points are $K_j^{(\infty)}$, then the algorithm produces

$$\mathbf{A_j^T} K_j^{(\infty)} + K_j^{(\infty)} \mathbf{A_j} + Q_j + \sum_{l=1, l \neq j}^{N} p_{lj} K_l^{(\infty)} = 0 \qquad (7.30)$$

Since (7.26) and (7.30) are identical, we conclude that Algorithm 7.6 converges to the unique solution of (7.26).   ∎

Note that if either $\left(\mathbf{A_j}, \sqrt{Q_j}\right)$ are observable or $Q_j > 0$ Algorithm 7.6 converges to the positive definite solutions $K_j > 0, \forall j$.

Example 7.2: In order to illustrate the procedure, consider a jump linear system having three Markovian transition states with transition rate matrix given by

$$p_{ij} = \{P\}_{ij}, \quad P = \begin{bmatrix} -3 & 2 & 1 \\ 1.5 & -2 & 0.5 \\ 0.75 & 0.75 & -1.5 \end{bmatrix}$$

The asymptotically stable system matrices are randomly generated and the matrices $Q_i, i = 1, 2, 3$, are taken as identities, that is

$$A_1 = \begin{bmatrix} -1.3232 & -1.1582 & 1.029 \\ -0.12292 & -2.0737 & 0.2234 \\ -0.6075 & 1.1656 & -3.1031 \end{bmatrix}, \quad Q_1 = I_3$$

$$A_2 = \begin{bmatrix} -2.479 & 1.3537 & -0.5717 \\ 0.8246 & -1.8727 & 0.4868 \\ 1.0958 & -0.9525 & -0.6483 \end{bmatrix}, \quad Q_2 = I_3$$

$$A_3 = \begin{bmatrix} -2.7604 & 0.5164 & -0.0381 \\ 0.5067 & -2.6064 & 0.399 \\ 0.528 & -0.2465 & -2.1332 \end{bmatrix}, \quad Q_3 = I_3$$

The obtained results are presented in the following table.

i	$\|K_1 - K_1^{(i)}\|_2$	$\|K_2 - K_2^{(i)}\|_2$	$\|K_3 - K_3^{(i)}\|_2$
1	0.065	0.078	0.0449
6	8.7822×10^{-4}	9.5615×10^{-4}	6.0326×10^{-4}
12	4.7078×10^{-6}	5.2211×10^{-6}	3.2586×10^{-6}
20	4.3339×10^{-9}	4.8151×10^{-9}	3.0006×10^{-9}

Table 7.2: Error propagation per iteration

It can be noticed that the algorithm is computationally very efficient since it took only 12 iterations to achieve accuracy of $O(10^{-6})$ for a system of 18 coupled scalar algebraic Lyapunov equations.

Δ

7.5.2 Discrete Coupled Algebraic Lyapunov Equations[2]

Discrete-time coupled algebraic Lyapunov equations corresponding to the discrete-time optimal jump linear control systems (Chizeck et al., 1986) are given by

$$K_j = A_j^T \left(\sum_{l=1}^{N} p_{jl} K_j \right) A_j + Q_j, \quad Q_j \geq 0, \ j = 1, 2, ..., N \quad (7.31)$$

where the subscript j indicates that the system is in mode j. In the discrete-time domain parameters p_{ij} satisfy $0 \leq p_{ij} \leq 1$ and $\sum_{i=1}^{N} p_{ij} = 1$, (Kumar and Varaiya, 1986). The following parallel algorithm has been proposed in (Borno and Gajic, 1995) for solving (7.31).

Algorithm 7.7:

$$K_j^{(m+1)} = A_j^T \left(\sum_{l=1}^{N} p_{jl} K_l^{(m)} \right) A_j + Q_j, \quad K_j^{(0)} = 0, \ j = 1, 2, ..., N$$

$$(7.32)$$

Δ

---

[2]   Reprinted with minor changes from *Computers & Mathematics with Applications,* vol.29, I. Borno and Z. Gajic, "Parallel algorithm for solving coupled algebraic Lyapunov equations of discrete-time jump linear systems," in press, 1995, with kind permission from Elsevier Science Ltd, The Boulevard, Langford Lane, Kidlington OX5 1GB, UK.

The convergence of this algorithm to the desired solution is obtained under the following assumptions.

Assumption 7.2 The system of coupled discrete algebraic Lyapunov equations (7.31) has unique positive semi-definite solutions $K_j, \forall j$.

$$\Delta$$

Assumption 7.3 Matrices $A_j, j = 1, 2, ..., N$, are asymptotically stable.

$$\Delta$$

Convergence Proof: Assumption 7.2 implies

$$A_j^T \left(\sum_{l=1}^{N} p_{jl} K_l \right) A_j \geq 0, \quad j = 1, 2, ..., N \quad (7.33)$$

For $m = 0$, Algorithm 7.7 produces

$$K_j^{(1)} = Q_j \geq 0, \quad j = 1, 2, ..., N \quad (7.34)$$

Hence,

$$0 \leq K_j^{(1)} \leq K_j, \quad \forall j \quad (7.35)$$

Similarly, it can be shown that

$$0 \leq K_j^{(1)} \leq K_j^{(2)} \leq K_j, \quad \forall j \quad (7.36)$$

By induction, it follows that the algorithm generates monotone nondecreasing sequences $\left\{ K_j^{(m)} \right\}$ bounded from above by the positive semi-definite solutions of (7.31), that is,

$$0 \leq K_j^{(1)} \leq K_j^{(2)} \leq ... \leq K_j^{(m)} \leq K_j^{(m+1)} \leq ... \leq K_j, \quad \forall j \quad (7.37)$$

Thus, the algorithm converges by the theory of monotonic convergence of positive operators (Kantorovich and Akilov, 1964). Note that Assumption 7.3 protects the solutions $K_j^{(m+1)}$ from escaping to infinity for large values of m. Substituting $m = \infty$ in (7.32) yields

$$K_j^{(\infty)} = A_j^T \left(\sum_{l=1}^{N} p_{jl} K_l^{(\infty)} \right) A_j + Q_j, \quad Q_j \geq 0, \quad j = 1, 2, ..., N \quad (7.38)$$

Since equations (7.31) and (7.38) are identical, the algorithm converges monotonically to the unique positive semi-definite solution of (7.31).

■

Example 7.3: Consider the following fourth-order jump linear system with three switching modes. The system matrices are randomly generated. The probability transition matrix is chosen as

$$P = \begin{bmatrix} 0.1 & 0.3 & 0.6 \\ 0.5 & 0.25 & 0.25 \\ 0 & 0.3 & 0.7 \end{bmatrix}$$

The problem matrices are

$$A_1 = \begin{bmatrix} 0.0667 & 0.0665 & 0.0844 & -0.2257 \\ 0.1383 & -0.1309 & 0.0797 & 0.1162 \\ 0.0658 & 0.0298 & 0.0645 & -0.1018 \\ -0.2283 & 0.2438 & -0.1990 & 0.2997 \end{bmatrix}, \quad Q_1 = I_4$$

$$A_2 = \begin{bmatrix} 01885 & -0.3930 & -0.0894 & -0.1919 \\ -0.4230 & 0.3598 & -0.1224 & -0.1548 \\ 0.0350 & -0.1950 & -0.1967 & -0.1017 \\ -0.2648 & -0.2440 & -0.0542 & 0.0484 \end{bmatrix}, \quad Q_2 = I_4$$

$$A_3 = \begin{bmatrix} 0.2746 & 0.0634 & 0.3414 & -0.0692 \\ 0.0796 & 0.4167 & 0.0283 & -0.1207 \\ -0.1607 & 0.0344 & -0.2227 & 0.1617 \\ 0.1175 & -0.2969 & 0.4149 & 0.3314 \end{bmatrix}, \quad Q_3 = I_4$$

The following solutions are obtained with accuracy of $O\left(10^{-12}\right)$ after 19 iterations

$$K_1 = \begin{bmatrix} 1.1013 & -0.0873 & 0.0826 & -0.0845 \\ -0.0873 & 1.1095 & -0.0644 & 0.0370 \\ 0.0826 & -0.0644 & 1.0705 & -0.0890 \\ -0.0845 & 0.0370 & -0.0890 & 1.2012 \end{bmatrix}$$

$$K_2 = \begin{bmatrix} 1.3727 & -0.3205 & 0.0423 & 0.0067 \\ -0.3205 & 1.4451 & 0.0404 & 0.0490 \\ 0.0423 & 0.0404 & 1.0733 & 0.0613 \\ 0.0067 & 0.0490 & 0.0613 & 1.0843 \end{bmatrix}$$

$$K_3 = \begin{bmatrix} 1.1271 & 0.0145 & 0.1859 & -0.0201 \\ 0.0145 & 1.3777 & -0.1309 & -0.2112 \\ 0.1859 & -0.1309 & 1.3609 & 0.0970 \\ -0.0201 & -0.2112 & 0.0970 & 1.2068 \end{bmatrix}$$

Since a system of 30 coupled scalar algebraic Lyapunov equations has been solved with high accuracy after only 19 iterations, we can observe that the presented algorithm is numerically very efficient. Simulation results are obtained by using MATLAB.

Define the error as

$$E^{(m)} = \max_j \left\| K_j^{(m)} - A_j^T \left(\sum_{l=1}^{N} K_l^{(m)} \right) A_j - Q_j \right\|_2 \qquad (7.39)$$

The error propagation per iteration is given in Table 7.3.

m	$E^{(m)}$
5	6.7956×10^{-4}
9	3.6998×10^{-6}
12	7.3945×10^{-8}
16	4.0100×10^{-10}
19	8.0134×10^{-12}

Table 7.3: Error propagation per iteration

Δ

7.6 Comments

Solutions of large algebraic Lyapunov equations are considered in (Hodel and Poolla, 1988, 1990), where the algorithms for computing the approximate solutions are presented. In (Gudmundsson and Laub, 1994) a method for estimating dominant eigenvalues and eigenvectors of the solution matrix P is given. This method also produces an approximate low-rank solution to the algebraic Lyapunov equation (7.1). The method is particularly efficient for algebraic Lyapunov equations having large and sparse coefficient matrices. The approximate low-rank solutions of the large algebraic Lyapunov equations are also obtained in (Saad, 1990), where the Krylov subspace method is used for solving (7.1) with $rank(Q) = 1$. The Krylov subspace method of (Saad, 1990) is extended in (Jaimoukha and Kasenally, 1994) for solving large algebraic Lyapunov equations with $rank(Q) > 1$. The paper approximates high order dimensional problem by low order problems, and shows how to get the approximate low-rank solutions. Both the continuous-time and discrete-time large algebraic Lyapunov equations are studied in (Jaimoukha and Kasenally, 1994).

The modified algebraic Lyapunov equation of the form

$$A^T P + PA + \mathcal{F}(P) + Q = 0 \qquad (7.40)$$

and linearly coupled algebraic Lyapunov equations

$$A_i^T P_i + P_i A_i + \sum_{j=1}^{N} \mathcal{F}_{ij}(P_j) + Q_i = 0, \quad i = 1, 2, ..., N \qquad (7.41)$$

where $\mathcal{F}$ and $\mathcal{F}_{ij}$ are linear matrix functions have been recently considered by (Richter et al., 1993). For the solution of the coupled algebraic Lyapunov equations (7.41), the reader is also referred to (Collins and Hodel, 1994).

7.7 References

1. Anderson, B., "Second-order convergent algorithms for the steady-state Riccati equation," *Int. J. Control*, vol.28, 295–306, 1978.

2. Balas, M., "Trends in large space structure control theory: Fondest hopes, wildest dreams," *IEEE Trans. Automatic Control*, vol.27, 522–535, 1982.

3. Barnett, S. and C. Storey, "Remarks on numerical solution of the Lyapunov matrix equation," *Electronic Letters*, vol.3, 417–418, 1967.

4. Barnett, S. and C. Storey, "Some applications of the Lyapunov matrix equation," *J. Inst. Math. Applics.*, vol.4, 33–42, 1968.

5. Borno, I., *Parallel Algorithms for Optimal Control of Linear Jump Parameter Systems and Markov Processes*, Doctoral Dissertation, Rutgers University, 1995a.

6. Borno, I. "Parallel computation of the solutions of coupled algebraic Lyapunov equations," *Automatica*, vol.31, in press, 1995b.

7. Borno, I. and Z. Gajic, "Parallel algorithm for solving coupled algebraic Lyapunov equations of discrete-time jump linear systems," *Computers & Mathematics with Appl.*, vol.29, in press, 1995.

8. Chizeck, H., A. Willsky, and D. Castanon, "Discrete-time Markovian jump linear quadratic optimal control," *Int. J. Control*, vol.13, 213–231, 1986.

9. Collins, E and A. Hodel, "Efficient solution of linearly coupled Lyapunov equations," *Proc. American Control Conference*, 2819–2823, Baltimore, 1994.

10. Davison, E. and F. Man, "The numerical solution of $A^T Q + QA = -C$," *IEEE Trans. Automatic Control*, vol.13, 448–449, 1968.

11. Ellner, N. and E. Wachspress, "Alternating direction implicit iteration for systems with complex spectra," *SIAM J. Numer. Anal.*, vol.28, 859–870, 1991.

12. Gudmundsson, T. and A. Laub, "Approximate solution to large sparse Lyapunov equations," *IEEE Trans. Automatic Control*, vol.39, 1110–1114, 1994.

13. Hodel, A., "The recent application of the Lyapunov equation in control theory," in *Iterative Methods in Linear Algebra*, 217–227, R. Beauwens and P. deGroen, Eds., North-Holland, Amsterdam, 1992.

14. Hodel, A. and K. Poolla, "Heuristic approaches to the solution of very large sparse Lyapunov and algebraic Riccati equations," *Proc. Decision and Control Conf.*, 2217–2222, Austin, 1988.

15. Hodel, A. and K. Poolla, "Numerical solution of very large, sparse Lyapunov equations through approximate power iterations," *Proc. Decision and Control Conf.*, 291–296, Honolulu, 1990.

16. Hodel, A. and K. Poolla, "Parallel solution of large Lyapunov equations," *SIAM J. Matrix Anal. Appl.*, vol.13, 1189–1203, 1992.

17. Jaimoukha, I. and E. Kasenally, "Krylov subspace methods for solving large Lyapunov equations," *SIAM J. Numer. Anal.*, vol.31, 227–251, 1994.

18. Ji, Y. and H. Chizeck, "Controllability, stabilizability, and continuous-time Markovian jump linear quadratic control, *IEEE Trans. Automatic Control*, vol.35, 777–788, 1990.

19. Jodar, L. and M. Mariton, "Explicit solutions for a system of coupled Lyapunov differential matrix equations," *Proc. Edinburgh Math. Soc.*, vol.30, 427–434, 1987.

20. Kantorovich, L. and G. Akilov, *Functional Analysis in Normed Spaces*. New York, Macmillan, 1964.

21. Kumar, P. and P. Varaiya, *Stochastic Systems: Estimation, Identification, and Adaptive Control*, Prentice Hall, Englewood Cliffs, 1986.

22. Lu, A. and E. Wachspress, "Solution of Lyapunov equations by alternating direction implicit iteration," *Computers & Mathematics with Appl.*, vol.21, 43–58, 1991.

23. Mariton, M., *Jump Linear Systems in Automatic Control*, New York — Basel, Marcel Dekker, 1990.

24. O'Leary and G. Stewart, "Data-flow algorithms for parallel matrix computations," *Comm. ACM.*, vol.28, 840–853, 1985.

25. Richter, S., L. Davis, and E. Collins, "Efficient computation of the solutions to modified Lyapunov equations," *SIAM J. Matrix Anal. Appl.*, vol.14, 420–431, 1993.

26. Smith, P., "Numerical solution of the matrix equation $AX + XA^T + B = 0$," *IEEE Trans. Automatic Control*, vol.16, 278-279, 1971.

27. Smith, R., "Matrix equation $XA + BX = C$," *SIAM J. Appl. Math.*, vol.16, 198-201, 1968.

28. Starke, G., "SOR for $AX + XB = C$," *Linear Algebra and Its Appl.*, vol.154–156, 355–375, 1991a.

29. Starke, G., "Optimal alternating direction implicit parameters for nonsymmetric systems of linear equations," *SIAM J. Numer. Anal.*, vol.28, 1431–1445, 1991b.

30. Starke, G., "SOR like methods for Lyapunov matrix equations," in *Iterative Methods in Linear Algebra*, 229–231, R. Beauwens and P. deGroen, Eds., North-Holland, Amsterdam, 1992.

31. Varga, R, *Matrix Iterative Analysis*, Prentice Hall, New York, 1962.

32. Wachspress, E., "Iterative solution of the Lyapunov matrix equation," *Appl. Math. Letters*, vol.1, 87–90, 1988.

33. Wachspress, E. "The ADI minimax problem for complex spectra," in *Iterative Methods for Large Linear Systems*, 251–271, Eds. D. Kincaid and L. Hayes, Academic Press, San Diego, 1990.

34. Wachspress, E., "ADI Iterative solution of Lyapunov equations," in *Iterative Methods in Linear Algebra*, 229–231, R. Beauwens and P. deGroen, Eds., North-Holland, Amsterdam, 1992.

35. Wonham, W., "Random difference equations in control theory," in *Probabilistic Methods in Applied Mathematics*, 131–212, Ed., A. Bharucha-Reid, Academic Press, New York, 1971.

36. Young, D., *Iterative Solution of Large Linear Systems*, Academic Press, New York, 1971.

Chapter Eight

Lyapunov Iterations

In this chapter we discuss applications of Lyapunov iterations as a powerful technique for solving some problems encountered in engineering and mathematics. The special attention is given to the Lyapunov iterations for finding solutions of several linear-quadratic optimal control problems such as linear regulator (filter), jump parameter linear systems, Nash differential games, and output feedback control.

The celebrated Kleinman algorithm for solving the main equation of the linear-quadratic optimal control theory, the Riccati equation, is presented in Section 8.1. The nonlinear algebraic Riccati equation is solved by performing iterations on linear Lyapunov algebraic equations. The convergence of the Kleinman algorithm to the positive semi-definite stabilizing solution is established. Examples are included to demonstrate the convergence of the presented algorithm to the both stabilizing solution (the closed-loop system matrix is asymptotically stable) and strong solution (the closed-loop system matrix is marginally stable).

In Section 8.2 we construct a sequence of Lyapunov algebraic equations whose solutions converge to the solutions of the coupled algebraic Riccati equations of the optimal control problem of jump parameter linear systems. The obtained solutions are positive semi-definite, stabilizing, and unique. The proposed algorithm is extremely efficient from the numerical point of view since it operates on the reduced-order *decoupled Lyapunov equations*.

In Section 8.3, the symmetric coupled algebraic Riccati equations corresponding to steady state Nash strategies are studied. Under control-oriented assumptions, imposed on the problem matrices, the Lyapunov iterations are constructed such that the proposed algorithm converges to the nonnegative (positive) definite stabilizing solution of the coupled algebraic Riccati equations. In addition, the problem order reduction is achieved since the obtained Lyapunov equations are of the reduced-order and can be solved independently. As a matter of fact, a parallel synchronous algorithm is obtained. A high-order numerical example is included in order to demonstrate the efficiency of the proposed algorithm.

In the remaining section of this chapter, we present the Lyapunov iterations for solving the output feedback control problem whose solution is given in terms of high order nonlinear algebraic equations (Moerder and Calise, 1985).

8.1 Kleinman's Algorithm for Riccati Equation

The linear-quadratic optimal control problem is defined by a controlled linear dynamic system

$$\dot{x} = Ax + Bu, \qquad x(t_0) = x_0 \tag{8.1}$$

and a quadratic performance criterion

$$J(u, x_0) = \frac{1}{2} \int_{t_0}^{\infty} \left(x^T Q x + u^T R u \right) dt, \qquad Q \geq 0, \ R > 0 \tag{8.2}$$

where $x \in \Re^n$ is a state vector, $u \in \Re^m$ is a control input. Matrices A, B, Q, and R are constant and of appropriate dimensions.

The optimal feedback solution to the optimization problem (8.1)-(8.2) is given by (Kwakernaak and Sivan, 1972)

$$u(t) = -R^{-1} B^T P x(t) \tag{8.3}$$

where P is the positive semi-definite stabilizing solution of the algebraic Riccati equation

$$A^T P + PA + Q - PSP = 0, \qquad S = BR^{-1}B^T \tag{8.4}$$

The required solution of (8.4) exists under the following standard control oriented assumption (Wonham, 1968).

Assumption 8.1 The triple $\left(A, B, \sqrt{Q}\right)$ is stabilizable-detectable.

$\triangle$

The unique positive semi-definite stabilizing solution of (8.4) is obtained in terms of the Lyapunov equations in (Kleinman, 1968). The Kleinman algorithm is given by

Algorithm 8.1:

$$\left(A - SP^{(i)}\right)^T P^{(i+1)} + P^{(i+1)}\left(A - SP^{(i)}\right) + Q + P^{(i)}SP^{(i)} = 0$$

$$with \ \left(A - SP^{(0)}\right) \ asymptotically \ stable$$

$$(8.5)$$

$\triangle$

In the following we show that under Assumption 8.1 any stabilizing initial guess $P^{(0)}$ makes Algorithm 8.1 convergent to the desired positive semi-definite stabilizing solution of (8.4).

Convergence Proof: From equations (8.4) and (8.5) we have

$$\left(A - SP^{(i)}\right)^T \left(P^{(i+1)} - P\right) + \left(P^{(i+1)} - P\right)\left(A - SP^{(i)}\right)$$

$$= -\left(P^{(i)} - P\right)S\left(P^{(i)} - P\right)$$

$$(8.6)$$

Since the right-hand side of this equation is negative semi-definite, it follows that if the unique solution of (8.6) exists, which is the case when $A - SP^{(i)} < 0$, $\forall i$, this solution must be positive semi-definite for every i, that is

$$P^{(i+1)} - P \geq 0, \quad i = 0, 1, 2, \dots \tag{8.7}$$

Also, it can be shown that from (8.5) we have

$$\left(A - SP^{(i+1)}\right)^T \left(P^{(i+2)} - P^{(i+1)}\right)$$

$$+ \left(P^{(i+2)} - P^{(i+1)}\right)\left(A - SP^{(i+1)}\right)$$

$$= \left(P^{(i)} - P^{(i+1)}\right)S\left(P^{(i)} - P^{(i+1)}\right), \quad i = 0, 1, 2, \dots$$

$$(8.8)$$

which by assuming asymptotic stability of matrices $A - SP^{(i+1)}$, $i = 0, 1, 2, ...$, implies

$$P^{(i+1)} \geq P^{(i+2)}, \quad i = 0, 1, 2, ... \qquad (8.9)$$

Thus, the solution matrices from Kleinman's Algorithm 8.1 form a monotonically nonincreasing sequence, which is bounded from bellow by the required solution of (8.4)

$$P^{(1)} \geq P^{(2)} \geq P^{(3)} \geq ... \geq P \qquad (8.10)$$

This sequence is convergent (Kantorovich and Akilov, 1964; Wonham, 1968).

It remains to establish the asymptotic stability of the sequence closed-loop matrices $A - SP^{(i+1)}$, $i = 0, 1, 2, ...$, assuming the initial matrix $A - SP^{(0)}$ is asymptotically stable. This is done in (Gohberg et al., 1986; Ran and Vreugdenhil, 1988) and stated here in the form of the following lemma.

Lemma 8.1 *The closed loop-system matrices $A - SP^{(i+1)}$ are asymptotically stable provided that the initial guess $P^{(0)}$ is stabilizing.*

□

Proof: The proof is done by using contradiction. Assume that $A - SP^{(i)}$ is a stable matrix, but the matrix $A - SP^{(i+1)}$ is unstable. Then, we have

$$\left(A - SP^{(i+1)}\right)v = \lambda v, \quad v \neq 0, \quad Re\{\lambda\} \geq 0 \qquad (8.11)$$

From (8.6) it is easy to derive

$$\left(A - SP^{(i+1)}\right)^T \left(P^{(i+1)} - P\right) + \left(P^{(i+1)} - P\right)\left(A - SP^{(i+1)}\right)$$
$$= -\left(P^{(i+1)} - P^{(i)}\right)S\left(P^{(i+1)} - P^{(i)}\right)$$
$$\quad -\left(P^{(i+1)} - P\right)S\left(P^{(i+1)} - P\right)$$

$$(8.12)$$

Using (8.11) in (8.12) implies

$$v^T 2Re\{\lambda\}\left(P^{(i+1)} - P\right)v =$$
$$-v^T\left(P^{(i+1)} - P^{(i)}\right)S\left(P^{(i+1)} - P^{(i)}\right)v \qquad (8.13)$$
$$-v^T\left(P^{(i+1)} - P\right)S\left(P^{(i+1)} - P\right)v = -v^T Mv$$

Since the left-hand side of the last equality is positive semi-definite by (8.7), and the right-hand side is negative semi-definite, it follows that $v^T M v = 0$ must be satisfied, which implies

$$S\left(P^{(i+1)} - P^{(i)}\right)v = 0 \qquad (8.14)$$

Thus, we have

$$\left(A - SP^{(i)}\right)v = \left(A - SP^{(i+1)}\right)v = \lambda v, \quad Re\{\lambda\} \geq 0 \qquad (8.15)$$

which contradicts the staring assumption that $\left(A - SP^{(i)}\right)$ is stable matrix and proves Lemma 8.1, (Gohberg, et al., 1986).

∎

In summary, established relations (8.7), (8.9)-(8.10), and Lemma 8.1, comprise the convergence proof of Kleinman's algorithm so that we have the following theorem.

∎

Theorem 8.1 *Under Assumption 8.1 the Kleinman algorithm (8.5) converges to the unique positive semi-definite stabilizing solution of the algebraic Riccati equation (8.4).*

□

Note that in the absence of detectability the Kleinman algorithm converges to the positive semi-definite solution of (8.4), which leads to the marginally stable closed-loop system matrix (eigenvalues are in the *closed* left side of the complex plane). This can be concluded from the work of (Poubelle et al., 1986). In Example 8.1 we have demonstrated both cases: asymptotically stable and marginally stable closed-loop matrices. The weaker assumption and the corresponding theorem are given below.

Assumption 8.2 The pair (A, B) is stabilizable.

△

Theorem 8.2 *Under Assumption 8.2 the Kleinman algorithm (8.5) converges to the strong solution (the closed loop-system matrix is marginally stable) of the algebraic Riccati equation (8.4).*

□

For completeness of the above analysis, let us mention that Assumption 8.1 can be replaced by the following assumption, which also guarantees the existence of the positive semi-definite stabilizing solution of the algebraic Riccati equation (8.4), (Kucera, 1972).

Assumption 8.3 The pair (A, B) is stabilizable and the Hamiltonian matrix, defined by

$$H = \begin{bmatrix} A & -S \\ -Q & -A^T \end{bmatrix}$$

has no pure imaginary eigenvalues.

△

Example 8.1 Consider the following linear-quadratic optimal control problem

$$A = \begin{bmatrix} -1 & 2 & 0 \\ 1 & 0 & 0 \\ 0 & 0 & 0 \end{bmatrix}, \ S = \begin{bmatrix} 0 & 0 & 0 \\ 0 & 3 & 0 \\ 0 & 0 & 3 \end{bmatrix}, \ Q = \begin{bmatrix} 1 & 0 & 0 \\ 0 & 0 & 0 \\ 0 & 0 & 0 \end{bmatrix}$$

Note that this system is only stabilizable and that the Hamiltonian matrix contains pure imaginary eigenvalues. The obtained solution is given by

$$P = \begin{bmatrix} 0.5856 & 0.5657 & 0 \\ 0 & 0.8685 & 0 \\ 0 & 0 & 0 \end{bmatrix} \geq 0$$

$$A - SP = \begin{bmatrix} -1 & 2 & 0 \\ -0.6972 & -2.6056 & 0 \\ 0 & 0 & 0 \end{bmatrix} \leq 0$$

that is, the closed-loop system matrix is only marginally stable.

If in the same example we change the matrix Q to

$$Q = diag\{0, \ 0, \ 1\}$$

it can be easily checked that there are no pure imaginary eigenvalues of the Hamiltonian matrix. The obtained solution is now stabilizable, that is, the closed-loop system matrix is asymptotically stable

$$P = \begin{bmatrix} 0.1667 & 0.3333 & 0 \\ 0.3333 & 0.6667 & 0 \\ 0 & 0 & 0.5774 \end{bmatrix} \geq 0, \quad \lambda(P) = \{0, 0.8333, 0.5774\}$$

$$A - SP = \begin{bmatrix} -1 & 2 & 0 \\ 0 & -2 & 0 \\ 0 & 0 & -1.7321 \end{bmatrix} < 0$$

$$\lambda(A - SP) = \{-1, -2, -1.7321\}$$

Δ

It is easy to show that the Kleinman algorithm is equivalent to the Newton method. Interestingly enough, it can be shown that the Kleinman algorithm can be obtained by using the successive approximations of dynamic programming (Bellman, 1954, 1957, 1961), and in the case of the algebraic Riccati equation this is equivalent to using the Newton method to solve it.

Note that "the differential form" of the Kleinman algorithm, that is, the differential Lyapunov iterations for solving the differential Riccati equation, was obtained in (Puri and Gruver, 1967).

Discrete-time Lyapunov iterations for solving the discrete-time nonlinear algebraic Riccati equation were presented in (Hewer, 1971; Kleinman, 1974).

8.2 Lyapunov Iterations for Jump Parameter Linear Systems[1]

Systems of coupled Riccati equations occur in several classes of optimal control problems such as jump linear control systems (Mariton, 1990). In this class of control problems a set of strongly coupled Riccati equations has to be solved in order to determine the optimal feedback gains. A

[1] © 1995 IEEE, Reprinted with permission from the paper "Lyapunov iterations for optimal control of jump linear systems at steady state," by Z. Gajic and I. Borno, *IEEE Trans. Automatic Control*, 1995, to appear.

homotopy algorithm for solving systems of coupled differential Riccati equations of jump parameter linear systems is presented in (Mariton and Bertrand, 1985). This method is computationally expensive and, hence, undesirable for solving algebraic equations of the corresponding steady state problem. In (Wonham, 1971) a method based on successive approximations led to the problem of solving a set of coupled differential Lyapunov equations. It can be easily observed that at steady state the method of (Wonham, 1971) is in fact the Newton method for solving the corresponding algebraic equations. The Newton algorithm, given in terms of the *coupled* algebraic Lyapunov equations, is also presented in (Salama and Gourishankar, 1974); however, while it is known for fast convergence, it suffers from its strong dependence on the proximity of the initial guess to the actual solution.

In this section, we present a new algorithm for solving coupled algebraic Riccati equations of jump parameter linear systems that converges to the optimal solution, regardless of the proximity of the initial guess to the actual solution, in a relatively small number of iterations. In addition, the algorithm is extremely efficient from the computational point of view since it operates only on the reduced-order *decoupled algebraic Lyapunov equations* (Gajic and Borno, 1995).

Consider a linear dynamic system described by

$$\dot{x}(t) = A(r)x(t) + B(r)u(t), \qquad x(t_0) = x_0 \qquad (8.16)$$

where $x(t)$ is an n-dimensional vector of system states, $u(t)$ is a control input of dimension m, A and B are mode-dependent matrices of appropriate dimensions, and r is a Markovian random process that represents the mode of the system and takes on values in a discrete set $\Psi = \{1, 2, ..., N\}$. The stationary transition probabilities of the modes of the system are determined by the transition rate matrix given by

$$\Pi = \begin{bmatrix} \pi_{11} & \pi_{12} & \cdots & \pi_{1N} \\ \pi_{21} & \pi_{22} & \cdots & \pi_{2N} \\ \vdots & \vdots & \vdots & \vdots \\ \pi_{N1} & \pi_{N2} & \cdots & \pi_{NN} \end{bmatrix} \qquad (8.17)$$

The matrix Π has the property that $\pi_{ij} \geq 0, i \neq j$, and $\pi_{ii} = -\sum_{j \neq i} \pi_{ij}$, (Varaiya and Kumar, 1986). The performance of the system (8.16) is evaluated by the criterion

$$J = E\left\{ \int\limits_0^\infty \left[x^T(t)Q(r)x(t) + u^T(t)R(r)u(t) \right] dt | t_0, x(t_0), r(t_0) \right\}$$

(8.18)

where $Q(r) \geq 0$ and $R(r) > 0$ for every r. The optimal feedback controls of (8.16)-(8.18) are given by, (Mariton, 1990)

$$u_{opt}(t) = -R_k^{-1}B_k^T P_k x(t), \quad k = 1, 2, ..., N$$

(8.19)

where the subscript k indicates that the system is in mode $r = k$, that is

$$A(r = k) = A_k, \quad B(r = k) = B_k$$
$$Q(r = k) = Q_k, \quad R(r = k) = R_k$$

(8.20)

and P_k, $k = 1, 2, ..., N$, are the positive semi-definite stabilizing solutions of a set of the coupled algebraic Riccati equations

$$\mathbf{A_k^T} P_k + P_k \mathbf{A_k} - P_k S_k P_k + Q_k + \sum_{j=1, j \neq k}^N \pi_{kj} P_j = 0, \quad k = 1, 2, ..., N$$

(8.21)

where

$$\mathbf{A_k} = A_k + \frac{1}{2}\pi_{kk} I, \quad S_k = B_k R_k^{-1} B_k^T$$

(8.22)

Equations (8.21)-(8.22) are nonlinear algebraic equations. The existence of positive semi-definite stabilizing solutions (stabilizable with respect to $\mathbf{A_k}$) of these equations is established in (Wonham, 1971) (see also Mariton, 1990; Wonham, 1968)) under the following assumption.

Assumption 8.4: The triples $\left(A_i, B_i, \sqrt{Q_i} \right)$, $i = 1, 2, , ..., N$, are stabilizable-detectable and

$$\max_{i=1,...,N} \left\{ \inf_{\Gamma_i} |\lambda_{max}[\int\limits_{t_0}^\infty e^{\left(A_i + B_i\Gamma_i + \frac{1}{2}\pi_{ii}I \right)^T t} \right.$$

(8.23)

$$\left. \times e^{\left(A_i + B_i\Gamma_i + \frac{1}{2}\pi_{ii}I \right)t} dt]| \right\} < 1$$

Δ

Condition (8.23) is a consequence of the fixed-point iterations (Wonham, 1968, 1971) used to establish the existence and uniqueness of the solutions of (8.21)-(8.22). It is "crucial ... and its conservativeness is difficult to evaluate", (Mariton, 1990). Note that in (Li and Chizeck, 1990) the notion of stochastic stabilizability is introduced to replace both the condition (8.23) and deterministic stabilizability of the pairs (A_i, B_i).

It is important to notice that Assumption 8.4 guarantees the stabilizability of the matrices $\mathbf{A_k}$ by the optimal closed-loop feedback gains. It is very natural to assume that the optimal feedback controls stabilize the actual systems, (Mariton, 1990), that is, to impose the additional assumption.

Assumption 8.5: The system matrices A_k, $k = 1, 2, ..., N$, are stabilizable by the optimal feedback controls (8.19).

$$\Delta$$

Assume that all conditions in Assumption 8.4 are satisfied, that is, assume that the unique stabilizing $P_k \geq 0$, $k = 1, 2, ..., N$, exist. The following algorithm (given in terms of decoupled algebraic Lyapunov equations) is proposed for solving the set of coupled algebraic Riccati equations (8.21)-(8.22), (Gajic and Borno, 1995).

Algorithm 8.2:

$$\left(\mathbf{A_k} - S_k P_k^{(i)}\right)^T P_k^{(i+1)} + P_k^{(i+1)}\left(\mathbf{A_k} - S_k P_k^{(i)}\right) =$$
$$-P_k^{(i)} S_k P_k^{(i)} - Q_k^{(i)}, \quad k = 1, 2, ..., N \quad (8.24)$$
$$with\ \ stabilizing\ \ P_k^{(0)} \geq 0$$

where

$$Q_k^{(i)} = Q_k + \sum_{j=1, j\neq k}^{N} \pi_{kj} P_j^{(i)} \geq 0 \quad (8.25)$$

$$\Delta$$

Thus, the solution of the $n \times N$-th order nonlinear coupled algebraic Riccati equations will be obtained by performing iterations on N decoupled linear algebraic Lyapunov equations each of order n.

Note that by (8.22) and Assumption 8.4, the triples $\left(\mathbf{A_k}, B_k, \sqrt{Q_k}\right)$ are stabilizable-detectable.

In the following, it will be shown that each sequence of solutions of (8.24)-(8.25) is nested between two sequences

$$K_k^{(i)} \le P_k^{(i)} \le \overline{P_k^{(i)}}, \qquad \forall i, \forall k \tag{8.26}$$

with $\left\{ K_k^{(i)} \right\}$ monotonically converging to P_k from below, that is

$$K_k^{(0)} \le K_k^{(1)} \le \dots \le P_k, \quad k = 1, 2, \dots, N \tag{8.27}$$

and the sequence $\left\{ \overline{P_k^{(i)}} \right\}$ monotonically converging to P_k from above, that is

$$\overline{P_k^{(0)}} \ge \overline{P_k^{(1)}} \ge \dots \ge P_k \tag{8.28}$$

Proof of Convergence:

Lower Bounds: Consider the following sequences of the standard algebraic Riccati equations

$$
\begin{aligned}
\mathbf{A_k^T} K_k^{(i+1)} + K_k^{(i+1)} \mathbf{A_k} - K_k^{(i+1)} S_k K_k^{(i+1)} + \mathbf{Q_k^{(i)}} &= 0 \\
\mathbf{Q_k^{(i)}} = Q_k + \sum_{j=1,j \ne k}^{N} \pi_{kj} K_j^{(i)}, \quad with \quad K_k^{(0)} &= 0, \forall k
\end{aligned} \tag{8.29}
$$

Note that by Assumption 8.4 the unique positive semi-definite stabilizing solutions of (8.29) exist for each iteration index i. For $i = 0$, we have

$$\mathbf{A_k^T} K_k^{(1)} + K_k^{(1)} \mathbf{A_k} - K_k^{(1)} S_k K_k^{(1)} + Q_k = 0, \quad k = 1, 2, \dots, N \tag{8.30}$$

By Assumption 8.4 the required positive semi-definite stabilizing solutions $P_k, k = 1, 2, \dots, N$, of (8.21) exist. Using the known results on comparison of the solutions for the standard algebraic Riccati equations, (see for example, Gohberg et al., 1986; Ran and Vreugdenhil, 1988), it follows from (8.21) and (8.30) that

$$Q_k \le Q_k + \sum_{j=1,j \ne k}^{N} \pi_{kj} P_j \implies P_k \ge K_k^{(1)} \tag{8.31}$$

For $i = 1$, we have

$$\mathbf{A_k^T} K_k^{(2)} + K_k^{(2)} \mathbf{A_k} - K_k^{(2)} S_k K_k^{(2)} + Q_k + \sum_{j=1,j\neq k}^{N} \pi_{kj} K_j^{(1)} = 0$$

$$k = 1, 2, ..., N$$

(8.32)

and since by (8.31)

$$Q_k + \sum_{j=1,j\neq k}^{N} \pi_{kj} K_j^{(1)} \leq Q_k + \sum_{j=1,j\neq k}^{N} \pi_{kj} P_j \qquad (8.33)$$

we get that $P_k \geq K_k^{(2)}$. Also, due to the fact

$$Q_k \leq Q_k + \sum_{j=1,j\neq k}^{N} \pi_{kj} K_j^{(1)} \qquad (8.34)$$

it follows $K_k^{(2)} \geq K_k^{(1)}$. Continuing the same procedure, we get from (8.29) monotonically nondecreasing sequences of positive semi-definite matrices bounded above by the solutions of (8.21), that is, by P_k

$$0 = K_k^{(0)} \leq K_k^{(1)} \leq K_k^{(2)} \leq ... \leq P_k, \quad k = 1, 2, ..., N \qquad (8.35)$$

These sequences are convergent and their limit points are $P_k, k = 1, 2, ..., N$, (Wonham, 1968; Kantorovich and Akilov, 1964).

Remark: Note that the recursive scheme (8.29), given in terms of the standard algebraic Riccati equations, can be used also for numerical solution of (8.21)-(8.22).

Now we show that the sequences of positive semi-definite matrices generated by the proposed algorithm, $\left\{ P_k^{(i)} \right\}$, are bounded from below by the sequences of positive semi-definite matrices generated by solving equations (8.29), that is, by $\left\{ K_k^{(i)} \right\}$. From equations (8.24)-(8.25) and

8.29) we get

$$\left(\mathbf{A_k} - S_k P_k^{(i)}\right)^T \left(P_k^{(i+1)} - K_k^{(i+1)}\right)$$
$$+ \left(P_k^{(i+1)} - K_k^{(i+1)}\right)\left(\mathbf{A_k} - S_k P_k^{(i)}\right) =$$
$$- \sum_{j=1, j\neq k}^{N} \pi_{kj}\left(P_j^{(i)} - K_j^{(i)}\right) - \left(P_k^{(i)} - K_k^{(i+1)}\right)S_k\left(P_k^{(i)} - K_k^{(i+1)}\right)$$
$$k = 1, 2, ..., N; \quad i = 1, 2, 3, ...$$

(8.36)

Since $K_k^{(0)} = 0$ and $P_k^{(0)} \geq 0$, $\forall k \Rightarrow P_k^{(0)} - K_k^{(0)} \geq 0$, it follows that the right-hand side of (8.36) is negative semi-definite so that for $i = 0$ we have $P_k^{(1)} - K_k^{(1)} \geq 0, \forall k$.

In order to establish that the matrices $\mathbf{A_k} - S_k P_k^{(i)}$ are stable for every k and i, we apply the stability proof technique from (Gohberg et al., 1986) to our problem. Rewrite (8.36) in the form

$$\left(\mathbf{A_k} - S_k P_k^{(i+1)}\right)^T \left(P_k^{(i+1)} - K_k^{(i+1)}\right)$$
$$+ \left(P_k^{(i+1)} - K_k^{(i+1)}\right)\left(\mathbf{A_k} - S_k P_k^{(i+1)}\right) =$$
$$= - \sum_{j=1, j\neq k}^{N} \pi_{kj}\left(P_j^{(i)} - K_j^{(i)}\right) - \left(P_k^{(i+1)} - P_k^{(i)}\right)S_k\left(P_k^{(i+1)} - P_k^{(i)}\right)$$
$$- \left(P_k^{(i+1)} - K_k^{(i+1)}\right)S_k\left(P_k^{(i+1)} - K_k^{(i+1)}\right)$$
$$k = 1, 2, ..., N; \quad i = 1, 2, 3, ...$$

(8.37)

The required stability proof technique from (Gohberg et al., 1986) is done by contradiction. Let us assume that the matrices $\mathbf{A_k} - S_k P_k^{(0)}$ are stable, but the matrix $\mathbf{A_k} - S_k P_k^{(1)}$ is unstable for some k. Then, there exists an eigenvalue λ such that

$$\exists k \ such \ that \ \left(A_k - S_k P_k^{(1)}\right)v = \lambda v, \ v \neq 0, \ Re\{\lambda\} \geq 0 \quad (8.38)$$

Using (8.38) in (8.37) we get

$$2v^T Re\{\lambda\}\left(P_k^{(1)} - K_k^{(1)}\right)v = -v^T M_k^{(1)}v, \quad M_k^{(1)} \geq 0, \ for \ some \ k$$

(8.39)

Since the left-hand side of (8.39) is positive semi-definite the equality in (8.39) is valid only for $v^T M_k^{(1)} v = 0$. From (8.37) we have

$$
\begin{aligned}
v^T M_k^{(1)} v = v^T [& \left(P_k^{(1)} - P_k^{(0)} \right) S_k \left(P_k^{(1)} - P_k^{(0)} \right) \\
& + \left(P_k^{(1)} - K_k^{(1)} \right) S_k \left(P_k^{(1)} - K_k^{(1)} \right) + \sum_{j=1, j \neq k}^{N} \left(P_j^{(0)} - K_j^{(0)} \right)] v
\end{aligned} \tag{8.40}
$$

which implies

$$
v^T \left(P_k^{(1)} - P_k^{(0)} \right) S_k \left(P_k^{(1)} - P_k^{(0)} \right) v = 0, \quad S_k \geq 0, \quad for \ some \ k \tag{8.41}
$$

or

$$
S_k \left(P_k^{(1)} - P_k^{(0)} \right) v = 0, \quad for \ some \ k \tag{8.42}
$$

Thus, we have obtained

$$
\left(\mathbf{A_k} - S_k P_k^{(0)} \right) v = \left(\mathbf{A_k} - S_k P_k^{(1)} \right) v = \lambda v, \quad v \neq 0, \quad for \ some \ k \tag{8.43}
$$

which is a contradiction due to the initial assumption that the matrices $\mathbf{A_k} - S_k P_k^{(0)}$ are stable for $\forall k$ so there is no such a k such that any of the matrices $\mathbf{A_k} - S_k P_k^{(1)}$ is unstable.

We have already established from (8.36) that $P_k^{(1)} - K_k^{(1)} \geq 0, \forall k$. Using this fact in (8.37) we see that the right-hand side of this equation is negative semi-definite, that is, $-x^T M^{(2)} x \leq 0$. Repeating steps (8.38)-(8.42) for $i = 1$, it follows that $\mathbf{A_k} - S_k P_k^{(2)}$ are stable matrices for $\forall k$. Continuing the same procedure for $i = 2, 3, ...$, we conclude that

$$
\begin{aligned}
\mathbf{A_k} - S_k P_k^{(i)} \ stable \ & \Rightarrow \ P_k^{(i+1)} - K_k^{(i+1)} \geq 0 \\
& \Rightarrow \ \mathbf{A_k} - S_k P_k^{(i+1)} \ stable \ \forall i, \forall k
\end{aligned} \tag{8.44}
$$

Thus, the sequences $\left\{ P_k^{(i)} \right\}$ are bounded from below by the sequences $\left\{ K_k^{(i)} \right\}, \forall k, \ \forall i$.

Upper Bounds: In the following we establish that the sequences $\left\{ P_k^{(i)} \right\}$ have the upper bounds, that is, the sequences $\left\{ \overline{P_k^{(i)}} \right\} \geq \left\{ P_k^{(i)} \right\}$

exist. In addition, these sequences (representing the upper bounds) monotonically converge from above to the required solutions of (8.21).

Subtracting (8.21) from (8.24) we get

$$\left(\mathbf{A_k} - S_k P_k^{(i)}\right)^T \left(P_k^{(i+1)} - P_k\right) + \left(P_k^{(i+1)} - P_k\right)\left(\mathbf{A_k} - S_k P_k^{(i)}\right)$$

$$= -\sum_{j=1, j\neq k}^{N} \pi_{kj}\left(P_j^{(i)} - P_j\right) - \left(P_k^{(i)} - P_k\right)S_k\left(P_k^{(i)} - P_k\right)$$

$$k = 1, 2, ..., N, \quad i = 0, 1, 2, ...$$

$$(8.45)$$

If for some iteration index i we have that

$$P_j^{(i)} - P_j \geq 0, \quad for \ \forall j \tag{8.46}$$

which can be obtained by choosing $P_j^{(0)} \geq P_j, \forall j$, then the right-hand side of (8.45) is negative semi-definite so that

$$P_k^{(i+1)} \geq P_k, \quad \forall k = 1, 2, ..., N, \quad \forall i = 0, 1, 2, ... \tag{8.47}$$

Even more, the sequences of matrices $\left\{P_k^{(i+1)}\right\}$ obtained from the corresponding algebraic Lyapunov equations are monotonically convergent with P_k representing their limit points. To show this, first observe from (8.24)-(8.25) that the following holds

$$\left(\mathbf{A_k} - S_k P_k^{(i+1)}\right)^T \left(P_k^{(i+2)} - P_k^{(i+1)}\right)$$

$$+ \left(P_k^{(i+2)} - P_k^{(i+1)}\right)\left(\mathbf{A_k} - S_k P_k^{(i+1)}\right) =$$

$$\sum_{j=1, j\neq k}^{N} \pi_{kj}\left(P_j^{(i)} - P_j^{(i+1)}\right) + \left(P_k^{(i)} - P_k^{(i+1)}\right)S_k\left(P_k^{(i)} - P_k^{(i+1)}\right)$$

$$k = 1, 2, ..., N; \quad i = 0, 1, 2, 3, ...$$

$$(8.48)$$

If in addition we impose

$$P_j^{(0)} - P_j^{(1)} \geq 0, \quad \forall j \tag{8.49}$$

then $P_k^{(i+2)} - P_k^{(i+1)} \leq 0, \forall i, \forall k$ so that monotonicity is obtained, that is

$$P_k^{(0)} \geq P_k^{(1)} \geq ... \geq P_k^{(i+1)} \geq P_k^{(i+2)} \geq ... \geq P_k \tag{8.50}$$

The sequences (8.50) are convergent, (Gohberg, 1986; Ran and Vreugdenhil, 1988; Wonham, 1968; Kantorovich and Akilov, 1964). Thus, the crucial point is the condition (8.49).

Consider now the sequences of positive semi-definite matrices generated by the proposed algorithm (8.24)-(8.25) with the stabilizing matrices $P_k^{(0)}$ taken "arbitrarily large" such that the condition (8.46) is satisfied. This can be always achieved by the stabilizability assumption. The required sequences are given by

$$
\left(\mathbf{A_k} - S_k \overline{P_k^{(i)}}\right)^T \overline{P_k^{(i+1)}} + \overline{P_k^{(i+1)}}\left(\mathbf{A_k} - S_k \overline{P_k^{(i)}}\right)
$$
$$
= -\overline{P_k^{(i)}} S_k \overline{P_k^{(i)}} - \overline{Q_k^{(i)}}, \qquad \overline{P_k^{(0)}} \geq P_k, \quad k = 1, 2, ..., N
$$
(8.51)

where

$$
\overline{Q_k^{(i)}} = Q_k + \sum_{j=1, j \neq k}^{N} \pi_{kj} \overline{P_j^{(i)}} \geq 0
$$
(8.52)

Note that the sequences $\left\{\overline{P_k^{(i)}}\right\}$ and $\left\{P_k^{(i)}\right\}$ are obtained from the same algorithm (8.24)-(8.25) and they only differ in the initial points $P_k^{(0)} \leq \overline{P_k^{(0)}}$. It can be easily shown by subtracting (8.51)-(8.52) from (8.24)-(8.25) that for any $0 \leq P_k^{(0)} \leq \overline{P_k^{(0)}}$ the sequences $\left\{P_k^{(i)}\right\}$ are dominated by the sequences $\left\{\overline{P_k^{(i)}}\right\}$, that is, the latter sequences represent the upper bounds for the former ones so that the corresponding inequalities (8.26) are satisfied. The sequences obtained, (8.51)-(8.52), must also satisfy the condition (8.49). For $i = 0$, we get from (8.51)

$$
\left(\mathbf{A_k} - S_k \overline{P_k^{(0)}}\right)^T \left(\overline{P_k^{(1)}} - \overline{P_k^{(0)}}\right) + \left(\overline{P_k^{(1)}} - \overline{P_k^{(0)}}\right)\left(\mathbf{A_k} - S_k \overline{P_k^{(0)}}\right)
$$
$$
= \overline{P_k^{(0)}} S_k \overline{P_k^{(0)}} - \overline{P_k^{(0)}} \mathbf{A_k} - \mathbf{A_k^T} \overline{P_k^{(0)}} - \overline{Q_k^{(0)}}, \quad k = 1, 2, ..., N
$$
(8.53)

Condition (8.49) will be satisfied if the right-hand side of (8.53) is positive semi-definite. It is shown in the next lemma that the right-hand side of (8.53) is positive semi-definite for $\overline{P_k^{(0)}} \geq P_k$, so that both conditions (8.46) and (8.49) are satisfied.

Lemma 8.2: *There exist $P_k^{(0)} \geq P_k, k = 1, 2, ..., N$, where P_k are unique positive semi-definite stabilizing solutions of (8.21)-(8.22), such that*

$$\Re\left(P_k^{(0)}\right) = P_k^{(0)} S_k P_k^{(0)} - \mathbf{A_k^T} P_k^{(0)} - P_k^{(0)} \mathbf{A_k} - Q_k$$

$$- \sum_{j=1, j \neq k}^{N} \pi_{kj} P_j^{(0)} \geq 0, \quad k = 1, 2, ..., N \tag{8.54}$$

$\square$

Proof: Let $P_k^{(0)} = P_k + X_k$ with $X_k \geq 0$, then

$$\Re\left(P_k^{(0)}\right) = \Re(P_k) + P_k S_k X_k + X_k S_k P_k + X_k S_k X_k$$

$$- A_k^T X_k - X_k A_k - \pi_{kk} X_k - \sum_{j=1, j \neq k}^{N} \pi_{kj} X_j \tag{8.55}$$

Using the fact that $\Re(P_k) = 0$, we rewrite (8.55) as

$$\Re\left(P_k^{(0)}\right) + \sum_{j=1}^{N} \pi_{kj} X_j - X_k S_k X_k = \tag{8.56}$$

$$-(A_k - S_k P_k)^T X_k - X_k (A_k - S_k P_k)$$

Since $X_k \geq 0$ and $A_k - S_k P_k$ are stable matrices for $\forall k$ by Assumption 8.5, it follows that the left-hand side of the Lyapunov equations (8.56) must be positive semi-definite, that is

$$\Re\left(P_k^{(0)}\right) + \sum_{j=1}^{N} \pi_{kj} X_j - X_k S_k X_k \geq 0 \tag{8.57}$$

Choosing $X_1 = X_2 = ... = X_k = ... = X_N$ and using the fact that $\pi_{kk} = - \sum_{j=1, j \neq k}^{N} \pi_{kj}$, it follows

$$\Re\left(P_k^{(0)}\right) \geq X_k S_k X_k \geq 0 \tag{8.58}$$

which completes the proof of Lemma 8.2.

■

Of course, the algorithm (8.51)-(8.52) by itself will converge to the desired solutions of (8.21)-(8.22), but very large values of $\overline{P_k^{(0)}}$ will slow the convergence process. Thus, the sequences obtained from (8.51) have only theoretical importance to establish the upper bounds for the sequences $\left\{ P_k^{(i)} \right\}$ since $0 \leq P_k^{(0)} \leq \overline{P_k^{(0)}}$, $\forall k$ imply $0 \leq P_k^{(i)} \leq \overline{P_k^{(i)}}$, $\forall i, \forall k$. The sequences $\left\{ P_k^{(i)} \right\}$ are used for actual computations. Note that the sequences $\left\{ P_k^{(i)} \right\}$ are nested between two sets of sequences

$$\left\{ K_k^{(i)} \right\} \leq \left\{ P_k^{(i)} \right\} \leq \left\{ \overline{P_k^{(i)}} \right\}, \quad \forall i, \forall k$$

Since both sequences $\left\{ K_k^{(i)} \right\}$ and $\left\{ \overline{P_k^{(i)}} \right\}$ converge to the required solutions of (8.21) so do the sequences $\left\{ P_k^{(i)} \right\}, \forall k$. Initial conditions for $\left\{ P_k^{(i)} \right\}, \forall k$ can be chosen as arbitrary positive semi-definite stabilizing matrices $P_k^{(0)} \geq 0$.

∎

Example 8.2: The following example was considered in (Mariton and Bertrand, 1985)

$$A_1 = diag(-2.5, -3, -2), \quad B_1 = diag\left(\sqrt{0.5}, 1, 1\right)$$
$$Q_1 = diag(25, 1, 11), \quad R_1 = I_3, \quad p_{11} = -3, \quad p_{12} = 0.5, \quad p_{13} = 2.5$$

$$A_2 = diag(-2.5, 5, 5), \quad B_2 = diag\left(\sqrt{0.5}, 1, \sqrt{0.5}\right)$$
$$Q_2 = diag(37.5, 704, 34.5), \quad R_2 = I_3, \quad p_{21} = p_{22} = p_{23} = 0$$

$$A_1 = diag(2, -3, -2), \quad B_3 = diag\left(\sqrt{0.5}, 1, 1\right)$$
$$Q_3 = diag(10, 16, 21), \quad R_3 = I_3, \quad p_{33} = p_{31} = p_{32} = 0$$

Using the initial conditions obtained from the decoupled algebraic Riccati equations (like in (8.30)), it took only 5 iterations for the proposed

Lyapunov iterations algorithm (8.24)-(8.25) to achieve the accuracy of $O\left(10^{-15}\right)$. Taking the initial guesses as $P_k^{(0)} = 100I_3, k = 1, 2, 3$, we got the accuracy of $O\left(10^{-15}\right)$ after ten iterations.

$\triangle$

Example 8.3: Consider the following fourth-order jump linear control problem

$$A_1 = \begin{bmatrix} -2.1051 & -1.1648 & 0.9347 & 0.5194 \\ -0.0807 & -2.8949 & 0.3835 & 0.8310 \\ 0.6914 & 10.5940 & -36.8199 & 3.8560 \\ 1.0692 & 13.4230 & 22.1185 & -13.1801 \end{bmatrix}$$

$$A_2 = \begin{bmatrix} -2.6430 & -1.2497 & 0.5269 & 0.6539 \\ -0.7910 & -2.8570 & 0.0920 & 0.4160 \\ 21.0357 & 22.8659 & -26.4655 & -1.7214 \\ 27.3096 & 7.8736 & -3.8604 & -29.5345 \end{bmatrix}$$

$$B_1 = \begin{bmatrix} 0.7564 \\ 0.9910 \\ 9.8255 \\ 7.2266 \end{bmatrix}, \ B_2 = \begin{bmatrix} 0.3653 \\ 0.2470 \\ 7.5336 \\ 6.5152 \end{bmatrix}, \ Q_1 = Q_2 = \begin{bmatrix} 1 & 0 & 1 & 0 \\ 0 & 0 & 0 & 0 \\ 1 & 0 & 1 & 0 \\ 0 & 0 & 0 & 0 \end{bmatrix}$$

$$\Pi = \begin{bmatrix} -2 & 2 \\ 1.5 & -1.5 \end{bmatrix}, \ R_1 = R_2 = 1$$

The following solutions have been obtained with the accuracy of $O\left(10^{-15}\right)$ after 14 iterations

$$P_1 = \begin{bmatrix} 0.2408 & 0.0705 & 0.0393 & 0.0182 \\ 0.0705 & 0.0308 & 0.0085 & 0.0064 \\ 0.0393 & 0.0085 & 0.0157 & 0.0025 \\ 0.0182 & 0.0064 & 0.0025 & 0.0016 \end{bmatrix}$$

$$P_2 = \begin{bmatrix} 0.5026 & 0.1343 & 0.0518 & 0.0097 \\ 0.1343 & 0.0485 & 0.0138 & 0.0026 \\ 0.0518 & 0.0138 & 0.0193 & 0.0002 \\ 0.0097 & 0.0026 & 0.0002 & 0.0003 \end{bmatrix}$$

The initial conditions for this problem were obtained by using solutions of the decoupled algebraic Riccati equations. Error propagation is given in Table 8.1, where the error is defined as

$$\max \left(\left\| \Re_1 \left(P_1^{(i)}, P_2^{(i)} \right) \right\|_2, \left\| \Re_2 \left(P_1^{(i)}, P_2^{(i)} \right) \right\|_2 \right)$$

All simulation results in this section are obtained by using MATLAB. It can be seen that the proposed algorithm has fast convergence, which in addition to the reduction in required computations and parallelism, makes this algorithm very attractive.

Iteration	Error
1	9.6000 x 10^{-2}
3	3.3579 x 10^{-4}
5	3.2379 x 10^{-6}
7	3.5113 x 10^{-8}
10	4.2811 x 10^{-11}
12	4.9626 x 10^{-13}
14	4.5259 x 10^{-15}

Table 8.1: Error propagation per iteration

$\triangle$

In the next two sections we show that the Lyapunov iterations are also efficient for solving the coupled algebraic Riccati equations of Nash differential games (having quadratic coupling) and highly nonlinear symmetric algebraic equations of the optimal output feedback control problem.

8.3 Lyapunov Iterations for Nash Differential Games

The solutions of the coupled algebraic Riccati equations produce the answers to some important problems of modern control theory, for example, differential games with conflict of interest and simultaneous decision making (Nash strategies), (Starr and Ho, 1969; Basar, 1991), H_∞ optimal control problems (Bernstein and Haddad, 1989; Basar and Bernhard, 1991), and jump linear systems (Mariton, 1990).

In this section, the solution of the symmetric coupled algebraic Riccati equations corresponding to steady state Nash strategies of the linear-quadratic differential game problem is presented in terms of the Lyapunov iterations by following work of (Gajic and Li, 1988; Li and Gajic, 1994). The obtained solution is stabilizing one, nonnegative (positive) definite and valid under the stabilizability-detectability assumptions imposed on the problem matrices. The proposed algorithm is of the reduced-order and can be implemented as a synchronous parallel algorithm (Bertsekas and Tsitsiklis, 1991). As a matter of fact, this algorithm is based on the successive approximations technique of dynamic programming (Bellman, 1954, 1957, 1961; Larson, 1967; Bertsekas, 1987). The method of successive approximations is the main tool in solving the functional equation of dynamic programming. It has been used in several control theory papers, for example (Vaisbord, 1963; Mil'shtein, 1964; Leake and Liu, 1967; Kleinman, 1968; Levine and Vilas, 1973; Mageriou, 1977). This method can be used as a very powerful decomposition technique which simplifies computations. In the work of (Mil'shtein, 1964), an approximate convergent method for synthesis of the optimal control system is investigated. The approach is based on a combination of the ideas of Lyapunov's second method and Bellman's method of successive approximations. Convergent suboptimal control sequences were also obtained in (Bellman, 1954, 1961) and (Vaisbord, 1963; Kleinman, 1968; Mageriou, 1977).

A controlled linear dynamic system corresponding to the Nash differential game strategies is given by

$$\dot{x} = Ax + B_1 u_1 + B_2 u_2, \qquad x(t_0) = x_0 \qquad (8.59)$$

where $x \in \Re^n$ is a state vector, $u_1 \in \Re^{m_1}$ and $u_2 \in \Re^{m_2}$ are control inputs (for the reason of simplicity we limit our attention to two control agents), A, B_1, and B_2 are constant matrices of appropriate dimensions.

With each control agent a quadratic type functional is associated, that is

$$J_1(u_1, u_2, x_0) = \frac{1}{2} \int_{t_0}^{\infty} \left(x^T Q_1 x + u_1^T R_{11} u_1 + u_2^T R_{12} u_2 \right) dt \quad (8.60)$$

$$J_2(u_1, u_2, x_0) = \frac{1}{2} \int_{t_0}^{\infty} \left(x^T Q_2 x + u_1^T R_{21} u_1 + u_2^T R_{22} u_2 \right) dt \quad (8.61)$$

Weighting matrices are symmetric and

$$Q_i \geq 0, \; i = 1, 2; \quad (positive \; semidefinite)$$
$$R_{ii} > 0, \; i = 1, 2; \quad (positive \; definite) \quad (8.62)$$
$$R_{ij} \geq 0, \; i = 1, 2; \; j = 1, 2; \; i \neq j$$

The optimal solution in the class of the best linear feedback laws forms the so-called Nash optimal strategies u_1^* and u_2^* satisfying

$$J_1(u_1^*, u_2^*) \leq J_1(u_1, u_2^*)$$
$$J_2(u_1^*, u_2^*) \leq J_1(u_1^*, u_2) \quad (8.63)$$

It was shown in (Starr and Ho, 1969) that the *closed-loop* Nash optimal strategy is given by

$$u_i^* = -R_{ii}^{-1} B_i^T K_i x, \quad i = 1, 2 \quad (8.64)$$

where K_i satisfy the coupled algebraic Riccati equations

$$K_1 A + A^T K_1 + Q_1 - K_1 S_1 K_1 - K_2 S_2 K_1 - K_1 S_2 K_2 + K_2 Z_2 K_2$$
$$= \mathcal{N}_1(K_1, K_2) = 0$$
$$\quad (8.65)$$
$$K_2 A + A^T K_2 + Q_2 - K_2 S_2 K_2 - K_2 S_1 K_1 - K_1 S_1 K_2 + K_1 Z_1 K_1$$
$$= \mathcal{N}_2(K_1, K_2) = 0$$
$$\quad (8.66)$$

with

$$S_i = B_i R_{ii}^{-1} B_i^T, \, i = 1, 2 \; ; \quad Z_i = B_i R_{ii}^{-1} R_{ji} R_{ii}^{-1} B_i^T, \, i, j = 1, 2; \; i \neq j \quad (8.67)$$

The existence of the nonlinear optimal Nash strategies was established in (Basar, 1974), so that (8.64), in fact, are the best linear optimal feedback strategies. Since a linear control law is highly desirable from the practical point of view, the linear feedback strategies (8.64) attract the attention of many researchers.

The existence of Nash strategies (8.64) and a solution of the coupled algebraic Riccati equations (8.65)-(8.66) were studied in (Papavassilopoulos et al., 1979) by means of the Brower fixed-point theorem by imposing norm conditions on the given matrices. These conditions are, in general, difficult to test and they are not very useful from a practical point of view. In the presented method, the control oriented assumptions (Wonham, 1968; Kucera, 1972) are imposed. It is shown that the obtained solution is nonnegative (positive) definite and stabilizing one. In addition, the proposed algorithm operates only on two decoupled standard algebraic Riccati equations (initialization) and performs iterations on two algebraic Lyapunov equations; thus, it operates on the reduced-order problems, and from a computational point of view, the algorithm is extremely efficient.

So far existing algorithms for solving (8.65)-(8.66), (Krikelis and Rekasius, 1971; Tabak, 1975), are of the local type, that is, they are faced with the problems of finding very good initial guesses. Furthermore, the algorithm proposed in (Tabak, 1975) does not necessarily converge even when the initial guesses are close to the optimal ones, as was pointed out in (Olsder, 1975). On the other hand, it is not known how to generate the initial guesses for the Newton-type algorithm used in (Krikelis and Rekasius, 1971), such that the algorithm converges to the stabilizing solutions of (8.65)-(8.66). Note that the differential coupled Riccati equations of Nash differential games corresponding to the finite time optimization problems were studied in (Jodar and Abou-Kandil, 1989; Abou-Kandil et al., 1993). In (Abou-Kandil et al., 1993) the nonsymmetric coupled algebraic Riccati equations corresponding to the *open-loop* Nash strategies were studied also. Equations (8.65)-(8.66) have been studied for special classes of systems in (Khalil and Kokotovic, 1979; Khalil, 1980) — singularly perturbed systems, and (Ozguner and Perkins, 1977; Petrovic and Gajic, 1988) — weakly coupled systems.

The considered algorithm is originally proposed by the authors in (Gajic and Li, 1988). It is shown that the algorithm converges to the nonnegative (positive) definite stabilizing solution of (8.65)-(8.66) under the following control-oriented assumption.

Assumption 8.6 Either the triple $\left(A, B_1, \sqrt{Q_1}\right)$ or $\left(A, B_2, \sqrt{Q_2}\right)$ is stabilizable-detectable.

$$\triangle$$

These conditions are quite natural since at least one control agent has to be able to control and observe unstable modes. Because the game is a noncooperative one, the assumption that their joint effect will take care of unstable modes seems to be very idealistic.

Let us suppose that $\left(A, B_1, \sqrt{Q_1}\right)$ is stabilizable-detectable. Then, a unique positive semi-definite solution of an auxiliary algebraic Riccati equation

$$K_1^{(0)} A + A^T K_1^{(0)} + Q_1 - K_1^{(0)} S_1 K_1^{(0)} = 0 \qquad (8.68)$$

exists such that $\left(A - S_1 K_1^{(0)}\right)$ is a stable matrix. By plugging $K_1 = K_1^{(0)}$ in (8.66) we get the second auxiliary Riccati equation as

$$K_2^{(0)} \left(A - S_1 K_1^{(0)}\right) + \left(A - S_1 K_1^{(0)}\right)^T K_2^{(0)} + \left(Q_2 + K_1^{(0)} Z_1 K_1^{(0)}\right)$$
$$- K_2^{(0)} S_2 K_2^{(0)} = 0$$
$$(8.69)$$

Since $\left(A - S_1 K_1^{(0)}\right)$ is a stable matrix and $Q_2 + K_1^{(0)} Z_1 K_1^{(0)}$ is a positive semi-definite matrix, the corresponding closed-loop matrix $\left(A - S_1 K_1^{(0)} - S_2 K_2^{(0)}\right)$ is stable. In fact, the triple $\left(A - S_1 K_1^{(0)}, B_2, \sqrt{Q_2 + K_1^{(0)} S_1 K_1^{(0)}}\right)$ is stabilizable-detectable and the stabilizing $K_2^{(0)}$ is uniquely determined. In the following we will use the solutions of (8.68)-(8.69), that is, $K_1^{(0)}$ and $K_2^{(0)}$ to initialize our algorithm.

The following algorithm is proposed in (Gajic and Li, 1988) for solving the coupled algebraic Riccati equations (8.65)-(8.66).

Algorithm 8.3:

$$\left(A - S_1 K_1^{(i)} - S_2 K_2^{(i)}\right)^T K_1^{(i+1)} + K_1^{(i+1)} \left(A - S_1 K_1^{(i)} - S_2 K_2^{(i)}\right) =$$
$$= \overline{Q_1^{(i)}} = -\left(Q_1 + K_1^{(i)} S_1 K_1^{(i)} + K_2^{(i)} Z_2 K_2^{(i)}\right), \quad i = 0, 1, 2, \dots$$
$$(8.70)$$

$$\left(A - S_1 K_1^{(i)} - S_2 K_2^{(i)}\right)^T K_2^{(i+1)} + K_2^{(i+1)}\left(A - S_1 K_1^{(i)} - S_2 K_2^{(i)}\right) =$$

$$= \overline{Q_2^{(i)}} = -\left(Q_2 + K_1^{(i)} Z_1 K_1^{(i)} + K_2^{(i)} S_2 K_2^{(i)}\right), \quad i = 0, 1, 2, \ldots$$

(8.71)

with initial conditions $K_1^{(0)}$ and $K_2^{(0)}$ obtained from (8.68)-(8.69).

$\triangle$

This algorithm is based on the Lyapunov iterations. Even though, it looks like that this algorithm has the form of (Kleinman, 1968), it is quite easy to show that this is not the case. Note that Kleinman's algorithm is used to solve the regular algebraic Riccati equation. Here, we are concerned with the problem of solving the coupled algebraic Riccati equations, where the coupling comes through the nonlinear quadratic terms, so that this problem is much more complex. Also, it can be shown that the proposed algorithm (8.70)-(8.71) is not of the Newton type.

The main property of algorithm (8.70)-(8.71) is given in the next theorem.

Theorem 8.3 *Under Assumption 8.6 the unique nonnegative definite stabilizing solution of the coupled algebraic Riccati equations (8.65)-(8.66) exists. It is obtained by performing Lyapunov iterations (8.70)-(8.71).*

□

The proof of this theorem is based on the successive approximations technique of dynamic programming, (Li and Gajic, 1994).

Note that the stronger result than the one stated in Theorem 8.3 can be similarly obtained by assuming that the penalty matrices Q_1 and Q_2 are positive definite.

Assumption 8.7 The state penalty matrices satisfy $Q_1 > 0$, $Q_2 > 0$.

$\triangle$

In that case we have the following theorem.

Theorem 8.4 *Under Assumptions 8.6 and 8.7 the unique positive definite stabilizing solution of the coupled algebraic Riccati equations (8.65)-(8.66) exists. It is obtained by performing Lyapunov iterations (8.70)-(8.71).*

□

Example 8.4:[2] In order to demonstrate the efficiency of the proposed algorithm we have run a tenth-order example, which is in fact a system of 110 nonlinear algebraic equations. Matrices A, B_1, and B_2 have been chosen randomly whereas the choice of matrices Q_1, Q_2, R_{11}, R_{12}, and R_{22} assures that Assumption 8.6 is satisfied. The corresponding matrices are given by

$$Q_1 = Q_2 = I_{10}$$

$$R_{11} = \begin{bmatrix} 1 & 0 \\ 0 & 2 \end{bmatrix}, \quad R_{12} = \begin{bmatrix} 3 & 0 \\ 0 & 4 \end{bmatrix}, \quad R_{21} = \begin{bmatrix} 5 & 0 \\ 0 & 6 \end{bmatrix}, \quad R_{22} = \begin{bmatrix} 7 & 0 \\ 0 & 8 \end{bmatrix}$$

$$A =$$

$$\begin{bmatrix}
-1.944 & 0.572 & 1.446 & -0.576 & 0.736 & -0.601 & -0.722 & -0.088 & 0.977 & 0.380 \\
1.440 & 0.393 & 1.023 & -0.711 & 1.282 & -0.679 & 0.010 & 0.588 & 1.281 & -1.414 \\
-0.881 & 1.058 & -1.492 & 1.113 & -1.728 & 0.498 & 0.313 & 1.509 & -1.536 & -0.264 \\
-1.170 & -1.055 & -0.058 & -0.723 & -0.939 & 1.453 & -1.087 & -0.486 & 1.066 & 0.235 \\
0.736 & -0.569 & 1.449 & -1.383 & 0.116 & -0.052 & 1.387 & 0.659 & -1.658 & -1.437 \\
0.014 & 0.658 & 0.586 & -0.850 & -0.074 & -1.335 & -0.261 & -1.021 & -0.449 & 1.444 \\
-0.734 & 0.621 & 0.422 & -0.369 & -0.395 & -0.453 & 1.228 & 0.213 & -1.380 & 1.307 \\
0.820 & -1.746 & 0.178 & -0.860 & -1.235 & -0.902 & 0.390 & -0.656 & -1.658 & 1.329 \\
0.831 & 0.569 & 1.408 & 1.500 & 1.396 & -0.605 & 0.387 & -0.729 & 1.717 & 1.309 \\
0.051 & -0.224 & 1.394 & 0.104 & -1.742 & -0.386 & -0.047 & -0.505 & -1.135 & 1.392
\end{bmatrix}$$

$$B_1 = \begin{bmatrix}
-2.036 & 0.637 \\
1.560 & 0.447 \\
-0.907 & 1.154 \\
-1.214 & -1.091 \\
0.813 & -0.575 \\
0.044 & 0.729 \\
-0.750 & 0.690 \\
0.901 & -1.826 \\
0.913 & 0.635 \\
0.084 & -0.209
\end{bmatrix}, \quad B_2 = \begin{bmatrix}
-1.648 & -0.759 \\
0.171 & 1.256 \\
-0.380 & -1.076 \\
-1.465 & -0.101 \\
1.854 & 0.745 \\
0.015 & 1.717 \\
0.458 & -0.091 \\
0.255 & -1.304 \\
0.274 & -0.763 \\
-0.502 & 1.345
\end{bmatrix}$$

The obtained results are really remarkable since only after 8 iterations we got very good convergence. These results are presented in Table 8.2. The errors are defined as the absolute values of the largest elements in matrices $\mathcal{N}_1\left(K_1^{(i)}, K_2^{(i)}\right)$ and $\mathcal{N}_2\left(K_1^{(i)}, K_2^{(i)}\right)$ where i stands for the number of iterations.

[2]   This example is reprinted with permission from the publisher from the book *Parallel Algorithms for Optimal Control of Large Scale Linear Systems*, by Z. Gajic and X. Shen, pp. 359–361, Springer Verlag, London, 1993.

Iteration	error 1	error 2
1	$1.5283 \times 10^{+2}$	$1.4193 \times 10^{+2}$
2	$1.6726 \times 10^{+1}$	$4.0585 \times 10^{+1}$
3	$3.1057 \times 10^{+0}$	$1.2188 \times 10^{+1}$
4	2.3207×10^{-1}	$2.4337 \times 10^{+0}$
5	1.2386×10^{-1}	7.6489×10^{-2}
6	4.1600×10^{-3}	6.9948×10^{-5}
7	7.0661×10^{-4}	3.0096×10^{-7}
8	2.4374×10^{-5}	9.2183×10^{-8}

Table 8.2: Simulation results for a
system of 110 nonlinear scalar equations

Simulation results are obtained by using the software package L-A-S
(Bingulac and Vanlandingham, 1993).

Δ

8.4 Lyapunov Iterations for Output Feedback Control

The linear-quadratic optimal output feedback control problem is defined
by

$$\dot{x} = Ax + Bu, \qquad x(t_0) = x_0 \qquad (8.72)$$

and

$$J(u, x_0) = \int_{t_0}^{\infty} \left(x^T Q x + u^T R u \right) dt, \qquad Q \geq 0, \; R > 0 \qquad (8.73)$$

where $x \in \Re^n$ is a state vector, $u \in \Re^m$ is a control input. Matrices $A, B, Q,$ and R are constant and of appropriate dimensions. *The* *minimizing control input $u(t)$ is constrained to*

$$u(t) = -Fy(t) \tag{8.74}$$

Many researchers studied the optimal control problem defined by (8.72)-(8.74) during the seventies and eighties. The analytical expression for the optimal constant output feedback gain F of the above constrained optimization problem was obtained by (Levine and Athans, 1970) as

$$F = R^{-1}B^T K L C^T \left(CLC^T\right)^{-1} \tag{8.75}$$

where the symmetric matrices K and L satisfy high-order coupled non-linear algebraic equations

$$(A - BFC)L + L(A - BFC)^T + x_0 x_0^T = 0 \tag{8.76}$$

$$(A - BFC)^T K + K(A - BFC) + Q + C^T F^T RFC = 0 \tag{8.77}$$

The required numerical solutions of high-order nonlinear algebraic equations (8.75)-(8.77) can be obtained by performing the Lyapunov iterations given in the next algorithm.

Algorithm 8.4:

Choose $F^{(0)}$ such that $A - BF^{(0)}C$ is a stable matrix $\tag{8.78}$

$$\left(A - BF^{(i)}C\right)L^{(i+1)} + L^{(i+1)}\left(A - BF^{(i)}C\right)^T + x_0 x_0^T = 0 \tag{8.79}$$

$$\left(A - BF^{(i)}C\right)^T K^{(i+1)} + K^{(i+1)}\left(A - BF^{(i)}C\right)^T + Q \\ + C^T F^{(i)^T} RF^{(i)}C = 0 \tag{8.80}$$

$$F^{(i+1)} = R^{-1}B^T K^{(i+1)} L^{(i+1)} C^T \left(CL^{(i+1)}C^T\right)^{-1} \tag{8.81}$$

$$\triangle$$

This algorithm converges to a local minimum of (8.73) under nonrestrictive assumptions (Moerder and Calise, 1985). Conditions for the convergence of Algorithm 8.4, the existence of a stabilizing solution, and the invertibility of $CL^{(i+1)}C^T$ term are discussed in (Moerder and

Calise, 1985) — see also (Toivonen, 1985). As a matter of fact, the updated value for F is defined in (Moerder and Calise, 1985) as

$$F_{new}^{(i+1)} = F^{(i)} + \alpha \left(F^{(i+1)} - F^{(i)} \right) \qquad (8.82)$$

where parameter $\alpha \in (0, 1]$ is chosen at each iteration such that the minimum is not overshot, that is

$$J^{(i+1)} = tr\left\{ K^{(i+1)} x_0^T x_0 \right\} < J^{(i)} = tr\left\{ K^{(i)} x_0^T x_0 \right\} \qquad (8.83)$$

Lyapunov iterations for solving the output feedback control problem of linear singularly perturbed systems are presented in (Gajic et al., 1989). In (Harkara et al., 1989) the Lyapunov iterations algorithm is developed for the output feedback control problem of linear weakly coupled systems.

8.4.1 Case Study: Fluid Catalytic Cracker

Algorithm 8.4 is demonstrated on the output feedback optimal control problem of a fluid catalytic cracker (Arkun and Ramakrishnan, 1983). The system matrix A is given in Example 2.7. The remaining matrices are given by

$$B^T = \begin{bmatrix} 11.12 & -3.61 & -21.91 & -53.6 & 69.1 \\ -12.6 & 3.36 & 0 & 0 & 0 \end{bmatrix}$$

$$C = \begin{bmatrix} 0 & 0 & 0 & 0 & 1 \\ 0 & 1 & 0 & 0 & 0 \end{bmatrix}, \quad Q = I_5, \quad R = I_2$$

Simulation results are presented in Table 8.3. The optimal value for the performance criterion is $J_{opt} = 0.28573$. It took 28 Lyapunov iterations to achieve the optimal result. The parameter α is chosen as $\alpha = 0.5$.

i	1	4	10	16	22	28
$J^{(i)}$	0.30487	0.28615	0.28583	0.28577	0.28574	0.28573

Table 8.3: Lyapunov iterations for the output feedback control

8.5 Comments

The method of Lyapunov iterations is a very powerful technique for approximation and decomposition of linear-quadratic optimal control problems and finding solutions of the corresponding nonlinear algebraic (and differential) equations. In addition to the presented methods, there are several other control problems where the Lyapunov iterations have been used to solve the nonlinear algebraic equations. For example, the Lyapunov iterations were used in the context of nonzero-sum differential games in (Petrovic and Gajic, 1988) for solving the coupled algebraic Riccati equations of Nash differential games for weakly coupled systems, in (Mageriou, 1977) for solving the algebraic Riccati equation of zero-sum differential games, and in (Mizukami and Suzumura, 1993) for finding the Stackelberg strategies for singularly perturbed systems.

8.6 References

1. Abou-Kandil, H., G. Freiling, and G. Jank, "Necessary conditions for constant solutions of coupled Riccati equations in Nash games," *Systems & Control Letters*, vol.21, 295–306, 1993.

2. Arkun, Y. and S. Ramakrishnan, "Bounds of the optimum quadratic cost of structure constrained regulators," *IEEE Trans. Automatic Control*, vol.28, 924–927, 1983.

3. Basar, T., "Generalized Riccati equations in dynamic games," in *The Riccati Equation*, Bittanti, S., A. Laub, and J. Willems, Eds., Springer-Verlag, London, 1991.

4. Basar, T. and P. Bernhard, H^∞— *Optimal Control and Related Minimax Design Problems: A Dynamic Game Approach*, Birhauser, Boston, 1991.

5. Basar, T., "A counterexample in linear-quadratic games: existence of non-linear Nash strategies," *J. Optimization Theory and Applications*, vol.14, 425–430, 1974.

6. Bellman, R., "Monotone approximation in dynamic programming and calculus of variations," *Proc. The National Academy of Science, USA*, vol.44, 1073–1075, 1954.

7. Bellman, R., *Dynamic Programming*, Princeton University Press, Princeton, 1957.

8. Bellman, R., *Adaptive Control Processes: A Guided Tour*, Princeton University Press, Princeton, 1961.

9.   Bernstein and W. Haddad, "LQG control with an H_∞ performance bound: A Riccati equation approach," *IEEE Trans. Automatic Control*, vol.34, 293–305, 1989.

10. Bertsekas, D., *Dynamic Programming: Deterministic and Stochastic Models*, Prentice Hall, Englewood Cliffs, 1987.

11. Bertsekas, D. and J. Tsitsiklis, "Some aspects of parallel and distributed iterative algorithms — A survey," *Automatica*, vol.27, 3–21, 1991.

12. Bingulac, S. and N. Vanlandingham, *Algorithms for Computer-Aided Design of Multivariable Control Systems*, Marcel Dekker, New York, 1993.

13. Gajic, Z. and T. Li, "Simulation results for two new algorithms for solving coupled algebraic Riccati equations," *Third Int. Symp. on Differential Games*, Sophia Antipolis, France, 1988.

14. Gajic, Z., D. Petkovski, and N. Harkara, "The recursive algorithm for the optimal static output feedback control problem of linear singularly perturbed systems," *IEEE Trans. Automatic Control*, vol.34, 465–468, 1989.

15. Gajic, Z. and I. Borno, "Lyapunov iterations for optimal control of jump linear systems at steady state," *IEEE Trans. Automatic Control*, to appear, 1995.

16. Gohberg, I., P. Lancaster, and L. Rodman, "On hermitian solutions of the symmetric algebraic Riccati equation," *SIAM J. Control and Optimization*, vol.24, 1323–1334, 1986.

17. Harkara, N., D. Petkovski, and Z. Gajic, "The recursive algorithm for the optimal static output feedback control of linear weakly coupled systems," *Int. J. Control*, vol.50, 1–11, 1989.

18. Hewer, G., "An iterative technique for the computation of the steady state gains for discrete optimal regulator," *IEEE Trans. Automatic Control*, vol.18, 382–284, 1971.

19. Ji, Y. and H. Chizeck, "Controllability, stabilizability, and continuous-time Markovian jump linear quadratic control," *IEEE Trans. Automatic Control*, vol.35, 777–788, 1990.

20. Jodar, L. and H. Abou-Kandil, "Kronecker products and coupled matrix Riccati differential equations," *Linear Algebra and Its Appl.*, vol.121, 39–51, 1989.

21. Kantorovich, L. and G. Akilov, *Functional Analysis in Normed Spaces*, Macmillan, New York, 1964.

22. Kleinman, D., "On an iterative technique for Riccati equation computation," *IEEE Trans. Automatic Control*, vol.13, 114–115, 1968.

23. Kleinman, D., "Stabilizing a discrete, constant, linear system with application to iterative methods for solving the Riccati equation," *IEEE Trans. Automatic Control*, vol.19, 253–254, 1974.

24. Khalil, H. and P. Kokotovic, "Feedback and well-posedness of singularly perturbed Nash games," *IEEE Trans. Automatic Control*, vol.24, 699–708, 1979.

25. Khalil, H., "Multimodel design of a Nash strategy," *J. Optimization Theory and Applications*, vol.31, 553–564, 1980.

26. Krikelis, N. and A. Rekasius, "On the solution of the optimal linear control problems under conflict of interest," *IEEE Trans. Automatic Control*, vol.16 140–147, 1971.

27. Kucera, V., "A contribution to matrix quadratic equations," *IEEE Trans. Automatic Control*, vol.17, 344–347, 1972.

28. Kwakernaak, H. and R. Sivan, *Linear Optimal Control Systems*, Wiley-Interscience, New York, 1972.

29. Larson, R., "A survey of dynamic programming computational procedures," *IEEE Trans. Automatic Control*, vol.12, 767–774, 1967.

30. Leake, R. and R. Liu, "Construction of suboptimal control sequences," *SIAM J. Control*, vol.5, 54–63, 1967.

31. Levine, M. and T. Vilas, "On-line learning optimal control using successive approximation techniques," *IEEE Trans. Automatic Control*, vol.19, 279–284, 1973.

32. Levine, W. and M. Athans, "On the determination of the optimal constant output feedback gains for linear multivariable systems," *IEEE Trans. Automatic Control*, vol.15, 44–48, 1970.

33. Li, T. and Z. Gajic, "Lyapunov iterations for solving coupled algebraic Lyapunov equations of Nash differential games and algebraic Riccati equations of zero-sum games," *Proc. Sixth Int. Symp on Dynamic Games and Appl.*, 489–494, St–Jovite, Canada, 1994.

34. Mageriou, E., "Iterative techniques for Riccati game equations," *J. Optimization Theory and Applications*, vol.22, 51–61, 1977.

35. Mariton, M., *Jump Linear Systems in Automatic Control*, Marcel Dekker, New York — Basel, 1990.

36. Mariton, M. and P. Bertrand, "A homotopy algorithm for solving coupled Riccati equations," *Optimal Control Appl. & Methods*, vol.6, 351–357, 1985.

37. Mil'shtein, G., "Successive approximation for solution of one optimum problem," *Automation and Remote Control*, vol.25, 298–306, 1964.

38. Mizukami, K. and F. Suzumura, "Closed-loop Stackelberg strategies for singularly perturbed systems: the recursive approach," *Int. J. Systems Science*, vol.24, 887–900, 1993.

39. Moerder, D. and A. Calise, "Convergence of numerical algorithm for calculating optimal output feedback gains," *IEEE Trans. Automatic Control*, vol.30, 900–903, 1985.

40. Olsder, G., "Comment on a numerical procedure for the solution of differential games," *IEEE Trans. Automatic Control*, vol.20, 704–705, 1975.

41. Ozguner, U. and W. Perkins, "A series solution to the Nash strategy for large scale interconnected systems," *Automatica*, vol.13, 313–315, 1977.

42. Papavassilopoulos, G., J. Medanic, and J. Cruz, "On the existence of Nash strategies and solutions to coupled Riccati equations in linear-quadratic games," *J. Optimization Theory and Appl.*, vol.28, 49–75, 1979.

43. Petrovic, B. and Z. Gajic, "The recursive solution of linear quadratic Nash games for weakly interconnected systems," *J. Optimization Theory and Appl.*, vol.56, 463–477, 1988.

44. Poubelle, M., I. Peterson, M. Gevers, and R. Bitmead, "A Miscellany of results on an equation of Count J. F. Riccati," *IEEE Trans. Automatic Control*, vol.31, 651–654, 1986.

45. Puri N. and W. Gruver, "Optimal control design via successive approximations," *Proc. Joint Automatic Control Conf.*, 335–344, Philadelphia, 1967.

46. Ran, A. and R. Vreugdenhil, "Existence and comparison theorems for algebraic Riccati equations for continuous- and discrete-time systems," *Linear Algebra and Its Appl.*, vol.99, 63–83, 1988.

47. Salama, A. and V. Gourishankar, "A computational algorithm for solving a system of coupled algebraic matrix Riccati equations," *IEEE Trans. on Computers*, vol.23, 100–102, 1974.

48. Starr, A. and Y. Ho., "Nonzero-sum differential games," *J. Optimization Theory and Appl.*, vol.3, 184–206, 1969.

49. Tabak, D., "Numerical solutions of differential game problems," *Int. J. Systems Science*, vol.6, 591–599, 1975.

50. Toivonen, H., "A globally convergent algorithm for the optimal constant output feedback problem," *Int. J. Control*, vol.41, 1589–1599, 1985.

51. Vaisbord, E., "An approximate method for the synthesis of optimal control," *Automation and Remote Control*, vol.24, 1626–1632, 1963.

52. Varaiya, P. and R. Kumar, *Stochastic Systems: Estimation, Identification, and Adaptive Control*, Prentice Hall, Englewood Cliffs, 1986.

53. Wonham, W., "On a matrix Riccati equation of stochastic control," *SIAM J. Control and Optimization*, vol.6, 681–697, 1968.

54. Wonham, W., "Random difference equations in control theory," 131–212, in *Probabilistic Methods in Applied Mathematics*, A. Bharucha-Reid, Ed., Academic Press, New York, 1971.

Chapter Nine

Concluding Remarks

In this concluding chapter we briefly discuss a Lyapunov-like equation, known as the Sylvester equation (Sylvester, 1884), fully present the list of applications of the Lyapunov and Lyapunov-like equations in science and engineering, and comment on the related areas not discussed in the previous chapters of this book. At the end of the chapter, in Section 9.4, we anticipate some interesting future research directions in the study of the Lyapunov equation.

9.1 Sylvester Equations

The Sylvester equation represents a generalization of the Lyapunov equation. The continuous-time algebraic Sylvester equation is defined by

$$AX + XB = C \tag{9.1}$$

where $A \in \Re^{m \times m}$, $B \in \Re^{n \times n}$, and $X, C \in \Re^{m \times n}$. *The existence of a unique solution of this equation is guaranteed if and only if the matrices A and $-B$ have no eigenvalues in common* (Gantmacher, 1959; Lancaster, 1970; Lancaster and Tismenetsky, 1985).

The discrete-time algebraic Sylvester equation given by

$$X - AXB = C \tag{9.2}$$

with the matrix dimensions as in (9.1), *has a unique solution provided that A and B have no mutually reciprocal eigenvalues* (Lancaster, 1970; Barnett, 1972). In addition, the unique solution of (9.2) can be represented by an infinite series

$$X = \sum_{i=1}^{\infty} A^{i-1} C B^{i-1} \tag{9.3}$$

if and only if $\rho(A)\rho(B) < 1$, where ρ represents the spectral radius. In (Smith, 1968) a technique is given to accelerate the convergence process of the series defined in (9.3). The corresponding convergence acceleration result of (Smith, 1968) for the algebraic Lyapunov equation is presented in Section 7.1 as Smith's algorithm.

In the work of (Gantmacher, 1959; Kucera, 1974), it has been shown that the general solution of the Sylvester equation (9.1) can be obtained as a sum of particular and homogeneous solutions, that is

$$X = X_p + X_h \tag{9.4}$$

where

$$\begin{aligned} AX_h + X_h B &= 0 \\ AX_p + X_p B &= C \end{aligned} \tag{9.5}$$

In addition, it is shown in (Kucera, 1974) that every solution of (9.1) has the form

$$X = VU^{-1} \tag{9.6}$$

with

$$U = [u_1, u_2, ..., u_n] \in \Re^{n \times n}, \quad V = [v_1, v_2..., v_n] \in \Re^{m \times n} \tag{9.7}$$

where $\left[u_i^T, v_i^T\right]^T$ represent the eigenvectors of the matrix

$$M = \begin{bmatrix} B & 0 \\ C & -A \end{bmatrix} \tag{9.8}$$

In view of the previous analysis, we have another existence result, which is established by (Roth, 1952; Kucera, 1974) and is given in the following theorem.

Theorem 9.1 *The Sylvester equation (9.1) has a solution if and only if the matrices*

$$\begin{bmatrix} B & 0 \\ C & -A \end{bmatrix} \quad and \quad \begin{bmatrix} B & 0 \\ 0 & -A \end{bmatrix} \qquad (9.9)$$

are similar.

☐

For the proof of Theorem 9.1 see (Kucera, 1974) and the corresponding correction from (Flanders and Wimmer, 1977). It can be also found in (Lancaster and Tismenetsky, 1985).

The Sylvester equation in the Jordan form was studied by (Rutherford, 1932) using an expansion method into a set of linear algebraic equations and in (Ma, 1966) by using a finite series method.

The Sylvester discrete-time algebraic equations with matrices A and B in companion canonical forms and with matrix C having all elements equal to zero except for $c_{mn} = 1$ is considered in (Bitmead, 1981). It is shown that the corresponding solution is a Toeplitz matrix.

Numerical solution of the algebraic Sylvester equation can be obtained by using the Bartels-Stewart algorithm, which is, as a matter of fact, originally developed for the Sylvester equation (Bartels and Stewart, 1972). Its specialized version for the algebraic Lyapunov equation is presented in Section 2.3.1. Also, the algorithm of (Golub et al., 1979) is applicable for the numerical solution of the Sylvester algebraic equation. Nowadays very popular Krylov subspace method is implemented in (Hu and Reichel, 1992) for solving the Sylvester algebraic equation, where it has been shown how to get the required solution of (9.1) in terms of several independent reduced-order subproblems. The method of (Hu and Reichel, 1992) extends the results of (Saad, 1990), where the Krylov subspace method is used for solving the large algebraic Lyapunov equations. The matrix sign function method for numerical solution of the continuous-time algebraic Sylvester equation is discussed in (Jodar, 1988).

Numerical solution of the differential Sylvester equation is considered in (Davison, 1975), and, in fact, it represents a generalization of Algorithm 4.1 presented in Chapter 4 for numerical solution of the differential Lyapunov equation. The algorithm of (Subrahmanyam, 1986), given in its specialized form in Section 4.3 for the differential Lyapunov equation is originally derived for the Sylvester differential equation. A time stepping procedure for direct numerical integration of the Sylvester differential equation is presented in (Serbin and Serbin, 1980). Coupled differential Sylvester equations are studied in (Jodar and Mariton, 1987).

Iterative methods presented in Chapter 7 for the Lyapunov equation are also applicable for the Sylvester equation (Smith, 1968; Wachspress, 1988; Starke, 1991, 1992). A parallel additive preconditioner for conjugate gradient method for solving the Sylvester algebraic equation is developed in (Evans and Galligani, 1994).

The algebraic Sylvester algebraic equation has been studied by many researchers in science and engineering, among them (Jameson, 1968; Muller, 1970; Jones and Lew, 1982).

A special type of the Sylvester equation is encountered in the estimator design. Given a linear time invariant system

$$\dot{x} = Ax + Bu$$
$$y = Cx \qquad\qquad (9.10)$$

Another linear time invariant dynamical system

$$\dot{z} = Fz + Gy + Hu \qquad\qquad (9.11)$$

is an estimator of the system (9.10) if $z(t) \rightarrow Tx(t)$, where T is a nonsingular matrix. The matrix T satisfies the following Sylvester equation (Chen, 1984)

$$-FT + TA = GC \qquad\qquad (9.12)$$

It is obvious from (9.11) that the matrix F must be asymptotically stable. Also, we already know, that for the existence of a unique solution of (9.12) the matrices A and F must not have eigenvalues in common. In addition to the above condition, *the existence of a unique invertible matrix T is guaranteed if and only if the pair (A, C) is observable and the pair (F, G) is controllable*, (Chen, 1984). More about the Sylvester equation coming from the estimator design can be found in (DeSouza and Bhattacharyya, 1981; B. Datta and K. Datta, 1992; B. Datta, 1994).

See also very interesting papers by (Tsui, 1987, 1993) on the analytical and numerical solutions of equation (9.12).

A special class of constrained Sylvester equations appear in robust observer design (Tsui, 1988) and loop transfer recovery problem (Barlow et al., 1992). The solution of (9.12) is required to satisfy

$$TB = 0, \quad \begin{bmatrix} T \\ C \end{bmatrix} \; has \; full \; rank \qquad (9.13)$$

Existence conditions for (9.12)-(9.13) together with an algorithm for finding T are given in (Barlow et al., 1992).

Constrained Sylvester equations corresponding to the problem of disturbance decoupling are presented in (Syrmos, 1994).

Existence conditions for a pair of generalized Sylvester equations of the form

$$A_1 X - Y B_1 = C_1$$
$$A_2 X - Y B_2 = C_2 \qquad (9.14)$$

are established in (Wimmer, 1994). Generalized Schur methods for solving generalized Sylvester equations (9.14) are given in (Kagstrom and Westin, 1989).

9.2 Related Topics

There are several additional topics on the Lyapunov equations which are not discussed in this book: periodic Lyapunov equations, existence of diagonal solutions, Lyapunov equation for characterization of positive real functions, and square root factor of the solution of the Lyapunov equation.

Periodic Lyapunov equations are considered in several paper, see for example, (Kwon and Pearson, 1982; Shayman, 1984; Bittanti et al., 1985; Bolzern and Colaneri, 1985; Bittanti and Colaneri, 1986) and references therein.

The problem of the existence of a positive diagonal matrix P such that $PA + A^T P < 0$ and a method for finding such a matrix is discussed in (Khalil, 1982). In (Geromel, 1985) a similar problem is considered for the continuous-time algebraic Lyapunov equation with requirement that both the matrices P and Q are diagonal. In the follow-up paper (Geromel and Santo, 1986), the problem is extended to the closed-loop matrix of a linear time invariant system that admits a diagonal positive solution for the Lyapunov equation.

The use of the algebraic Lyapunov equation in characterizing the positive real matrices is presented in (Anderson, 1967; Dickinson, 1980). This problem is related to the well known "positive real lemma" of Kalman and Yakubovich. Discrete-time relation of the Kalman-Yakubovich lemma and the Lyapunov (Sylvester) equation is given in (Bitmead, 1981).

In (Larin and Aliev, 1993) a method of finding a square root of the solution of the continuous-time algebraic Lyapunov equation, without actually solving the equation, is obtained by using the matrix sign function method. Having obtained the square root the solution of the considered Lyapunov equation is easily constructed. Even more, the square root of the solution of the Lyapunov equation can be used in calculating the balancing transformation (Laub et al., 1987). Note that the algorithm of Hammarling, based on the Schur decomposition, in the process of solving the algebraic Lyapunov equation first finds the square root (Cholesky factor) of the solution matrix and then the actual solution (Hammarling, 1982).

Of course, the Lyapunov-like equations (Sylvester equations) are just marginally touched in this book. *Another book can be written on the Sylvester equation* in order to complement the work on the most important linear matrix equations in science and engineering.

9.3 Applications

Main applications of the Lyapunov and Lyapunov-like equations are outlined in Chapter 1. Here, we expand the list of applications of these equations and give the full account of engineering and scientific disciplines where the Lyapunov/Lyapunov-like equations are present.

The Lyapunov type matrix equations appear in many diverse engineering and mathematics perspectives such as control theory, system theory, optimization, power systems, signal processing, linear algebra, differential equations, boundary value problems, large space flexible structures, bilinear systems, and communications.

The Lyapunov matrix equation is not only encountered in studying the stability of linear systems (Halanay and Rasvan, 1993), but also in other areas of *system/control theory and practice*. It has been presented in Section 1.3 that the quadratic performance measure of a linear feedback system can be evaluated in terms of the solution of the Lyapunov equation (Kwakernaak and Sivan, 1972). Also, we have seen in Section 1.2 that for stochastic linear systems driven by white noise, the solution of the Lyapunov equation represents the variance of the state vector (Kwakernaak and Sivan, 1972).

Many other control and system problems are based on the Lyapunov and/or Lyapunov-like equations such as: controllability and observability grammians (Chen, 1984; Bender, 1987), balancing transformation (Moore, 1981; Laub et al., 1987; Therapos, 1989), stability robustness to parameter variations (Patel and Toda, 1980; Yedavalli, 1985; Zhou and Khargonekar, 1987), robust stability and performance study of large scale systems (Hyland and Bernstein, 1987), reduced-order modeling and control (Hyland and Bernstein, 1985, 1986; Bernstein and Hyland, 1985; Safonov and Chiang, 1989), filtering with singular measurement noise (Haddad and Bernstein, 1987; Halevi, 1989), filtering error due to inaccuracy in the filter's initial conditions (Boka and Gajic, 1992), estimator design (Chen, 1984; Tsui, 1987, 1988, 1993). The Lyapunov and/or Lyapunov-like equations also appear in differential games for solving the algebraic Riccati equation of zero-sum games (Mageirou, 1977), Nash strategies for weakly coupled systems (Petrovic and Gajic, 1988), Stackelberg strategies (Mizukami and Suzumura, 1993), singular systems (Bender, 1987; Lewis and Mertzios, 1987; Lewis and Ozcaldiran, 1989; Owens and Debeljkovic, 1986), estimate of root location of linear systems (Mori and Kuwahara, 1982), stability of differential systems with delay (Agathoklis and Foda, 1989), output feedback control (Moerder and Calise, 1985; Toivonen, 1985; Bernstein, 1987), loop transfer recovery (Barlow et al., 1992), analysis of bilinear systems (Lewis et al., 1990), computation of the $\mathcal{H}_2$ norm of stable transfer functions (Doyle et al., 1989), and disturbance decoupling problems (Syrmos, 1994). Also, the definition of stochastic stabilizability is given in terms of the coupled algebraic Lyapunov equations (Ji and Chizeck, 1990).

In *mathematics* the Lyapunov-like equations were subject of research since the beginning of this century (Wedderburn, 1904). A method of undetermined coefficients for solving systems of linear differential equations, based on a Lyapunov-like equation, is presented in (Dou, 1966). The Lyapunov/Lyapunov-like equations also appear in boundary value problems in partial differential equations (Kreisselmeier, 1972), stability analysis of interval matrices (Foo and Soh, 1990), characterization of positive real matrices (Anderson, 1967; Dickinson, 1980), numerical solution of implicit differential equations (Epton, 1980), solution of the discrete Poisson equation (Dorr, 1970), and interpolation problems of rational matrix functions (Lerer and Rodman, 1993). A sequences of generalized Lyapunov equations can be used for polynomial matrix factorization with respect to the imaginary axis (Aliev and Larin, 1993).

The Lyapunov/Lyapunov-like equations have been recently used in *digital signal processing* (Anderson et al., 1986; Agathoklis, 1988; Lu et al., 1992), construction of canonical ladder forms for vector autoregressive processes (Hadidi et al., 1982), stability study of 2–dimensional discrete systems (Tzafestas et al., 1992; Agathoklis et al., 1993), for Gohberg-Semencul inversion formulas for Hermitian Toeplitz and quasi Toeplitz matrices appearing in signal processing (Pal, 1993), stability analysis of 2–dimensional digital filters (Hinamoto, 1993; Lu, 1994), and generation of discrete-time q-Markov covers (Sreeram et al., 1994; Sreeram, 1994).

In addition, the Lyapunov type equations are encountered in identification and network synthesis (Mehra, 1970; Bitmead, 1981), power systems (Ilic, 1989; Hodel, 1992), beam gridworks problem in *mechanical engineering* (Ma, 1966), large space flexible structures (Balas, 1982), stability of the second-order matrix polynomials appearing in classical mechanics, aerodynamic, and robotics (Shieh et al., 1987).

In the next subsections we demonstrate two useful applications of the algebraic Lyapunov equation in the problems of finding the approximate performance criteria in linear-quadratic control systems and determining the variance of the state variables of linear systems under white noise disturbances.

9.3.1 Case Study: Magnetic Tape Control System

Differential equation describing the control problem of a magnetic tape is given in (Chow and Kokotovic, 1976)

$$\dot{x}(t) = Ax(t) + Bu(t) = \begin{bmatrix} 0 & 0.4 & 0 & 0 \\ 0 & 0 & 0.345 & 0 \\ 0 & -5.24 & -4.65 & 2.62 \\ 0 & 0 & 0 & -10 \end{bmatrix} x(t) + \begin{bmatrix} 0 \\ 0 \\ 0 \\ 10 \end{bmatrix} u(t)$$

In control engineering the goal is to find a control input $u(t)$ such that a quadratic performance criterion

$$J = \frac{1}{2} \int\limits_0^\infty \left[x^T(\tau) R_1 x(\tau) + u^T(\tau) R_2 u(\tau) \right] d\tau$$

is minimized. The penalty matrices R_1 and R_2 for this particular problem are chosen as (Chow and Kokotovic, 1976)

$$R_1 = diag\{1, 0, 1, 0\}, \quad R_2 = 1$$

The optimal control law that produces the global minimum is given in terms of the solution of the algebraic Riccati equation, see Section 8.1. If one uses an approximate linear feedback control in the form $u(t) = Fx(t)$, then from the results presented in Section 1.3 we have

$$J_{app} = \frac{1}{2} x^T(0) P x(0)$$

where P satisfies the algebraic Lyapunov equation

$$(A - BF)^T P + P(A - BF) + R_1 + F^T R_2 F = 0$$

Even more, if F stabilizes $(A - BF)$, then according to the Kleinman algorithm from Section 8.1, the following sequence of the algebraic Lyapunov equations

$$\left(A - BF^{(i)} \right)^T P^{(i+1)} + P^{(i+1)} \left(A - BF^{(i)} \right) + R_1 + F^{T(i)} R_2 F^{(i)} = 0$$

$$F^{(0)} = F, \quad F^{(i)} = R_2^{-1} B^T P^{(i)}, \quad i = 1, 2, \dots$$

will produce the minimal (optimal) value for the performance criterion, that is

$$J_{app}^{(i)} = 0.5x^T(0)P^{(i)}x(0) \to J_{opt} = J_{min}$$

The convergence rate of the above sequence is quadratic since, in fact, it represents the Newton method for solving the algebraic Riccati equation. Using the initial value for the feedback gain matrix as $F = \begin{bmatrix} 1 & 1 & 1 & 1 \end{bmatrix}$, the sequence of the approximate values that converge to the minimal (optimal) value is obtained and presented in Table 9.1. The initial condition for this problem is taken as $x(0) = \begin{bmatrix} 1 & 1 & 1 & 1 \end{bmatrix}^T$, and the optimal value of the performance criterion is given by $J_{opt} = 14.7565$.

i	$J_{app}^{(i)}$
1	18.9434
2	15.2613
3	14.7658
4	14.7565

Table 9.1: Approximate performance criteria
and their convergence to the optimal one

9.3.2 Case Study: Aircraft under Wind Disturbances

Consider the following forth-order model of an F-8 aircraft studied in (Teneketzis and Sandell, 1977; Khalil and Gajic, 1984). The aircraft dynamics under wind disturbances is described by

$$\dot{x}(t) = Ax(t) + \Gamma w(t)$$

$$= \begin{bmatrix} -0.01357 & -32.2 & -46.3 & 0 \\ 0.00012 & 0 & 1.214 & 0 \\ -0.00012 & 0 & -1.214 & 1 \\ 0.00057 & 0 & -9.010 & -0.6696 \end{bmatrix} x(t) + \begin{bmatrix} -46.3 \\ 1.214 \\ -1.214 \\ -9.01 \end{bmatrix} w(t)$$

where $w(t)$ is a white noise Gaussian stochastic process representing the wind disturbance. It is known that the wind can be quite accurately modelled as a white noise stochastic process with the intensity matrix V, where $E\{w(t)w^T(\tau)\} = V\delta(t-\tau)$ and δ is the delta impulse function. The variance of the state variables at steady state $(t \rightarrow \infty)$ for this system is given by the solution of the following algebraic Lyapunov equation (see Section 1.2)

$$AP + PA^T + \Gamma V\Gamma^T = 0, \quad Var(x(t)) = P$$

Using MATLAB and its function lyap and the value for the wind intensity as $V = 0.000315$, (Teneketzis and Sandell, 1977) the following result is obtained for the variance of the state variables of the considered aircraft

$$Var(x(t)) = P = \begin{bmatrix} 0.4731 & -0.0050 & 0.0106 & 0.0326 \\ -0.0050 & 0.0001 & -0.0002 & -0.0009 \\ 0.0106 & -0.0002 & 0.0009 & 0.0009 \\ 0.0326 & -0.0009 & 0.0009 & 0.0068 \end{bmatrix}$$

It can be noticed that the variable x_1 is the most affected by the wind since its variance of 0.4731 is much bigger than the variances of the remaining state variables.

9.4 Comments

The Lyapunov equation is a simple linear equation, but extremely reaching in its applications. Due to its importance many journal papers have been already written (more than 250), but the Lyapunov equation is still not fully explored. In the years to come, we expect to see the papers clarifying the best possible bounds for the trace, determinant, and eigenvalues of the solution of the Lyapunov equation, continuation of the development of parallel and iterative algorithms for solving large scale Lyapunov equations, theoretical studies of coupled Lyapunov equations and unified continuous-discrete-time Lyapunov equations.

Algebraic Lyapunov equations having nonunique solutions appear in the context of singular systems. Due to the lack of results available in the literature these equations are not treated in this book. However, they might be an interesting subject for future research. Another interesting problem of the symmetric Lyapunov equations, corresponding to the linear systems having symmetric state matrices (many linearized systems in mechanical engineering have this form), should be also considered by the researchers interested in the Lyapunov equation and its applications.

9.5 References

1. Agathoklis, P., "The Lyapunov equation for n-dimensional discrete systems," *IEEE Trans. Automatic Control*, vol.35, 448–451, 1988.

2. Agathoklis, P. and S. Foda, "Stability and the matrix Lyapunov equation for delay differential systems," *Int. J. Control*, vol.49, 417–432, 1989.

3. Agathoklis, P., E. Jury, and M. Mansour, "Algebraic necessary and sufficient conditions for the stability of 2–D discrete systems," *IEEE Trans. Circuits and Systems — II: Analog and Digital Signal Processing*, vol.40, 251–258, 1993.

4. Aliev, F. and V. Larin, "Generalized Lyapunov equation and factorization of matrix polynomials," *Systems & Control Letters*, vol.21, 485–491, 1993.

5. Anderson, B., "A system theory criterion for positive real matrices," *SIAM J. Control*, vol.5, 171–182, 1967.

6. Anderson, B., P. Agathoklis, E. Jury, and M. Mansour, "Stability and the matrix Lyapunov equation for discrete 2–dimensional systems," *IEEE Trans. Circuits and Systems*, vol.33, 261–266, 1986.

7. Balas, M., "Trends in large space structure control theory: Fondest hopes, wildest dreams," *IEEE Trans. Automatic Control*, vol.27, 522–535, 1982.

8. Barlow, J., M. Monahem, and D. O'Leary, "Constrained matrix Sylvester equations," *SIAM J. Matrix Anal. Appl.*, vol.13, 1–9, 1992.

9. Barnett, S., *Matrices in Control Theory*, van Nostrand-Reinhold, London, 1972.

10. Bartels, R. and G. Stewart, "Algorithm 432, solution of the matrix equation $AX + XB = C$," *Comm. Ass. Computer Machinery*, vol.15, 820-826, 1972.

11. Bender, D., "Lyapunov-like equations and reachability/observability Gramians for descriptor systems," *IEEE Trans. Automatic Control*, vol.32, 343–348, 1987.

12. Bernstein, D., "The optimal projection equations for static and dynamic output feedback: the singular case," *IEEE Trans. Automatic Control*, vol.32, 1139–1143, 1987.

13. Bernstein, D. and D. Hyland, "The optimal projection equation for reduced-order state estimation," *IEEE Trans. Automatic Control*, vol.30, 583–585, 1985.

14. Bitmead, R., "Explicit solutions of the discrete-time Lyapunov matrix equation and Kalman-Yakubovich equations," *IEEE Trans. Automatic Control*, vol.26, 1291–1294, 1981.

15. Bittanti, S., P. Bolzen, and P. Colaneri, "The extended periodic Lyapunov lemma," *Automatica*, vol.21, 603–605, 1985.

16. Bittanti, S. and P. Colaneri, "Lyapunov and Riccati equations: Periodic inertia theorems," *IEEE Trans. Automatic Control*, vol.31, 659–661, 1986.

17. Boka, J. and Z. Gajic, "Kalman filtering error due to inaccuracy in the filter initial condition," *Proc. IEEE Regional Control Conf.*, 134–136, New York, 1992.

18. Bolzern, P. and P. Colaneri, "Inertia theorems for the periodic Lyapunov difference equation and periodic Riccati difference equation," *Linear Algebra and Its Appl.*, vol.85, 247–265, 1987.

19. Chen, C., *Linear System Theory and Design*, Holt, Rinehart and Winston, New York, 1984.

20. Chow, J. and P. Kokotovic, "A decomposition of near-optimum regulators for systems with slow and fast modes," *IEEE Trans. Automatic Control*, vol.21, 701–705, 1976.

21. Datta, B., "Linear and numerical linear algebra in control theory: Some research problems," *Linear Algebra and Its Appl.*, vol.197–198, 755–790, 1994.

22. Datta, B. and K. Datta, "High performance computing in linear control," *Proc. The 12th World Congress of IFAC*, vol.9, 493–500, Sydney, Australia, 1993.

23. Davison, E., "The numerical solution of $\dot{X} = A_1 X + X A_2 + D, \quad X(0) = C$," *IEEE Trans. Automatic Control*, vol.20, 566-567, 1975.

24. DeSouza, E. and S. Bhattacharyya, "Controllability, observability and the solution of $AX - XB = C$," *Linear Algebra and Its Appl.*, vol.39, 167–188, 1981.

25. Dickinson, B., "Analysis of the Lyapunov equation using generalized positive real matrices," *IEEE Trans. Automatic Control*, vol.25, 560–563, 1980.

26. Dorr, F., "The direct solution of the discrete Poisson equation on a rectangle," *SIAM Review*, vol.12, 248–263, 1970.

27. Dou, A., "Method of undetermined coefficients in linear differential systems and the matrix equation $YB - AY = F$," *SIAM J. Appl. Math.*, vol.14, 691–696, 1966.

28. Doyle, J., K. Glover, P. Khargonekar, and B. Francis, "State-space solutions to standard H_2 and H_∞ control problems," *IEEE Trans. Automatic Control*, vol.34, 831–847, 1989.

29. Epton, M., "Methods for the solution of $AXD - BXC = E$ and its application in the numerical solution of implicit ordinary differential equations," *BIT*, vol.20, 341–345, 1980.

30. Evans, D. and E. Galligani, "A parallel additive preconditioner for conjugate gradient method for $AX + XB = C$," *Parallel Computing*, vol.20, 1055–1064, 1994.

31. Flanders, H. and H. Wimmer, "On the matrix equations $AX - XB = C$ and $AX - YB = C$," *SIAM J. Appl. Math.*, vol.32, 707–710, 1977.

32. Foo, Y. and Y. Soh, "Stability analysis of a family of matrices," *IEEE Trans. Automatic Control*, vol.35, 1257–1259, 1990.

33. Gantmacher, F., *The Theory of Matrices*, vol.1 and 2, Chelsea, New York, 1959.

34. Geromel, J., "On the determination of a diagonal solution of the Lyapunov equation," *IEEE Trans. Automatic Control*, vol.30, 404–406, 1985.

35. Geromel, J. and A. Santo, "On the robustness of linear continuous-time dynamic systems," *IEEE Trans. Automatic Control*, vol.31, 1136–1138, 1986.

36. Golub, G., S. Nash, and C. Loan, "A Hessenberg-Schur method for the problem $AX + XB = C$," *IEEE Trans. Automatic Control*, vol.24, 909–913, 1979.

37. Haddad, W. and D. Bernstein, "The optimal projection equations for reduced-order state estimation: the singular measurement noise," *IEEE Trans. Automatic Control*, vol.32, 1135-1143, 1987.

38. Hadidi, M., M. Morf, and B. Porat, "Efficient construction of canonical ladder forms for vector autoregressive processes," *IEEE Trans. Automatic Control*, vol.27, 1222–1232, 1982.

39. Halanay, A. and V. Rasvan, *Applications of Lyapunov Methods in Stability*, Kluwer, Dordrecht, The Netherlands, 1993.

40. Halevi, Y., "The optimal reduced-order estimator for systems with singular measurement noise," *IEEE Trans. Automatic Control*, vol.34, 777–781, 1989.

41. Hammarling, S., "Numerical solution of the stable, non-negative definite Lyapunov equation," *IMA J. Numerical Analysis*, vol.2, 303–323, 1982.

42. Hinamoto, T., "2–D Lyapunov equation and filter design based on Fornasini-Marchesini second method," *IEEE Trans. Circuits and Systems — I: Fundamental Theory and Applications*, vol.40, 102–110, 1993.

43. Hodel, A., "The recent application of the Lyapunov equation in control theory," in *Iterative Methods in Linear Algebra*, 217–227, R. Beauwens and P. DeGroen, Eds., North-Holland, Amsterdam, 1992.

44. Hyland, D. and D. Bernstein, "The optimal projection equations for model reduction and the relationships among the methods of Wilson, Skelton, and Moore," *IEEE Trans. Automatic Control*, vol.30, 1201–1211, 1985.

45. Hyland, D. and D. Bernstein, "The optimal projection equations for finite dimensional fixed-order dynamic compensation of infinite-dimensional systems," *SIAM J. Control and Optimization*, vol.24, 122–151, 1986.

46. Hyland, D. and D. Bernstein, "The majorant Lyapunov equation: A nonnegative matrix equation for robust stability and performance of large scale systems," *IEEE Trans. Automatic Control*, vol.32, 1005–1013, 1987.

47. Hu, D. and L. Reichel, "Krylov-subspace methods for the Sylvester equation," *Linear Algebra and Its Appl.*, vol.172, 283–313, 1992.

48. Ilic, M., "New approaches to voltage monitoring and control," *IEEE Control System Magazine*, vol.9, 3–11, 1989.

49. Jameson, A., "Solution of the equation $AX + XB = C$ by inversion of $M \times M$ and $N \times N$ matrix," *SIAM J. Appl. Math.*, vol.16, 1020–1023, 1968.

50. Ji, Y. and H. Chizeck, "Controllability, stabilizability, and continuous-time Markovian jump linear quadratic control," *IEEE Trans. Automatic Control*, vol.35, 777–788, 1990.

51. Jodar, L., "An algorithm for solving generalized algebraic Lyapunov equations in Hilbert space, applications to boundary value problems," *Proc. Edinburgh Math. Soc.*, vol.31, 99–105, 1988.

52. Jodar, L. and M. Mariton, "Explicit solutions for a system of coupled Lyapunov differential matrix equations," *Proc. Edinburgh Math. Soc.*, vol.30, 427–434, 1987.

53. Jones, J. and C. Lew, "Solutions of the Lyapunov matrix equation $BX - XA = C$," *IEEE Trans. Automatic Control*, vol.27, 464–466, 1982.

54. Kagstrom, B. and L. Westin, "Generalized Schur methods with condition estimators for solving the generalized Sylvester equation," *IEEE Trans. Automatic Control*, vol.34, 745–751, 1989.

55. Khalil, H., "On the existence of positive diagonal P such that $PA + A^T P < 0$," *IEEE Trans. Automatic Control*, vol.27, 181–184, 1982.

56. Khalil, H. and Z. Gajic, "Near-optimum regulators for stochastic linear singularly perturbed systems," *IEEE Trans. Automatic Control*, vol.29, 531–541, 1984.

57. Kreisselmeier, G., "A solution of the bilinear matrix equation $AY + YB = -Q$," *SIAM J. Appl. Math.*, vol.23, 334–338, 1972.

58. Kucera, V., "The matrix equation $AX + XB = C$," *SIAM J. Appl. Math.,* vol.26, 15–25, 1974.

59. Kwakernaak, H. and R. Sivan, *Linear Optimal Control Systems,* Wiley, 1972.

60. Kwon, W. and A. Pearson, "Linear systems with two-point boundary Lyapunov and Riccati equations," *IEEE Trans. Automatic Control,* vol.27, 436–441, 1982.

61. Lancaster, P., "Explicit solutions of linear matrix equations," *SIAM Review,* vol.12, 544–566, 1970.

62. Lancaster, P. and M. Tismenetsky, *Theory of Matrices,* Academic Press, New York, 1985.

63. Larin, V. and F. Aliev, "Construction of square root factor for solution of the Lyapunov matrix equation," *Systems & Control Letters,* vol.20, 109–112, 1993.

64. Laub, A., M. Heath, C. Paige, and R. Ward, "Computation of system balancing transformations and other applications of simultaneous diagonalization algorithms," *IEEE Trans. Automatic Control,* vol.32, 115–122, 1987.

65. Lerer, L. and L. Rodman, "Sylvester and Lyapunov equations and some interpolation problems for rational matrix functions," *Linear Algebra and Its Appl.,* vol.185, 83–117, 1993.

66. Lewis, F. and V. Mertzios, "Analysis of singular systems using orthogonal functions," *IEEE Trans. Automatic Control,* vol.32, 527–530, 1987.

67. Lewis, F., V. Mertzios, G. Vachtsevanos, and M. Christodoulou, "Analysis of bilinear systems using Walsh functions," *IEEE Trans. Automatic Control,* vol.35, 119–123, 1990.

68. Lewis, F. and K. Ozcaldiran, "Geometric structure and feedback in singular systems," *IEEE Trans. Automatic Control,* vol.34, 450–455, 1989.

69. Lu, W., "On a Lyapunov approach to stability analysis of 2–D digital filters," *IEEE Trans. on Circuits and Systems — I: Fundamental Theory and Applications,* vol.41, 665–669, 1994.

70. Lu, W., H., Wang, and A. Antoniou, "An efficient method for evaluation of the controllability and observability gramians of 2–D digital filters and systems," *IEEE Trans. on Circuits and Systems,* vol.39, 695–704, 1992.

71. Ma, E., "A finite series solution of the matrix equation $AX - XB = C$," *SIAM J. Appl. Math.,* vol.14, 490–495, 1966.

72. Mageriou, E., "Iterative techniques for Riccati game equations," *J. Optimization Theory and Appl.,* vol.22, 51–61, 1977.

73. Mehra, R., "An algorithm to solve matrix equations $PH^T = G$ and $P = \Phi P \Phi^T + \Gamma \Gamma^T$," *IEEE Trans. Automatic Control,* vol.15, 600, 1970.

74. Mizukami, K. and F. Suzumura, "Closed-loop Stackelberg strategies for singularly perturbed systems: the recursive approach," *Int. J. Systems Science,* vol.24, 887–900, 1993.

75. Moerder, D. and A. Calise, "Convergence of numerical algorithm for calculating optimal output feedback gains," *IEEE Trans. Automatic Control,* vol.30, 900–903, 1985.

76. Moore, B., "Principal component analysis in linear systems: Controllability, observability and model reduction," *IEEE Trans. Automatic Control,* vol.26, 17–32, 1981.

77. Mori, T. and M. Kuwahara, "Estimate for the root-location of linear systems via the Lyapunov matrix equation," *J. Franklin Institute,* vol.314, 123–127, 1982.

78. Muller, P., "Solution of the matrix equations $AX + XB = -Q$ and $S^T X + XS = -Q$," *SIAM J., Appl. Math.,* vol.18, 682–687, 1970.

79. Owens, D. and D. Debeljkovic, "Consistency and Liapunov stability of linear descriptor systems: A geometric analysis," *IMA J. Mathematical Control and Information,* vol.2, 139–151, 1986.

80. Pal., D., "Gohberg-Secencul type formulas via embedding of Lyapunov equations," *IEEE Trans. Signal Processing,* vol.41, 2208–2215, 1993.

81. Patel, R. and M. Toda, "Quantitative measures of robustness for multivariable systems," *Proc. Joint American Control Conf.,* paper TP8–A, San Francisco, 1980.

82. Petrovic, B. and Z. Gajic, "The recursive solution of linear quadratic Nash games for weakly interconnected systems," *J. Optimization Theory and Appl.,* vol.56, 463-477, 1988.

83. Roth, W., "The equations $AX - YB = C$ and $AX - XB = C$ in matrices," *Proc. Amer. Math. Soc.,* vol.3, 392–396, 1952.

84. Rutherford, D., "On the solution of the matrix equation $AX + XB = C$," *Nederl. Akad. Wenesch. Proc., Series A.,* vol.35, 53–59, 1932.

85. Saad, Y., "Numerical solution of large Lyapunov equations," in *Signal Processing, Scattering, Operator Theory and Numerical Methods,* J. Kaashoek, H. van Schuppen, and A. Rao, Eds., Birkhauser, Boston, 1990.

86. Safonov, M. and R. Chiang, "A Schur method for balanced-truncation model reduction," *IEEE Trans. Automatic Control,* vol.34, 729–733, 1989.

87. Serbin, S. and C. Serbin, "A time-stepping procedure for $\dot{X} = A_1 X + XA_2 + D,\ X(0) = C$," *IEEE Trans. Automatic Control,* vol.25, 1138–1141, 1980.

88. Shayman, M., "Inertia theorems for the periodic Lyapunov equation and periodic Riccati equation," *Systems & Control Letters,* vol.4, 27–32, 1984.

89. Shieh, L., M. Mehio, and H. Dib, "Stability of the second-order matrix polynomial, *IEEE Trans. Automatic Control,* vol.32, 231–233, 1987.

90. Smith, R., "Matrix equation $XA + BX = C$," *SIAM J. Appl. Math.,* vol.16, 198-201, 1968.

91. Sreeram, V., "On the generalized q-Markov cover models for discrete-time systems," *IEEE Trans. Automatic Control,* vol.39, 2502–2505, 1994.

92. Sreeram, V., P. Agathoklis, and M. Mansour, "The generation of discrete-time q-Markov covers via inverse solution of the Lyapunov equation," *IEEE Trans. Automatic Control,* vol.39, 381–385, 1994.

93. Starke, G., "SOR for $AX - XB = C$," *Linear Algebra and Its Appl.,* vol.154–156, 355–375, 1991.

94. Starke, G., "SOR-like methods for Lyapunov matrix equations," in *Iterative Methods in Linear Algebra,* 233–240, R. Beauwens and P. deGroen, Eds., North-Holland, Amsterdam, 1992.

95. Subrahmanyam, M., "On a numerical method of solving the Lyapunov and Sylvester equations," *Int. J. Control,* vol.43, 433-439, 1986.

96. Sylvester, J., "Sur la solution du cas le plus général des équations linéaires en quantités binaires, c̀est-à-dire en quaternions ou en matrices du second ordre", *C. R. Acad. Sci. Paris,* vol.22, 117–118, 1884.

97. Syrmos, V., "Disturbance decoupling using constrained Sylvester equations," *IEEE Trans. Automatic Control,* vol.39, 797–803, 1994.

98. Teneketzis, D. and N. Sandell, "Linear regulator design for stochastic systems by multiple time-scale method," *IEEE Trans. Automatic Control,* vol.22, 615–621, 1977.

99. Therapos, C., "Balancing transformations for unstable nonminimal linear systems," *IEEE Trans. Automatic Control,* vol.34, 455–457, 1989.

100. Toivonen, H., "A globally convergent algorithm for the optimal constant output feedback problem," *Int. J. Control,* vol.41, 1589–1599, 1985.

101. Tsui, C., "A complete analytical solution to the equation $TA - FT = LC$ and its applications," *IEEE Trans. Automatic Control,* vol.32, 742–744, 1987.

102. Tsui, C., "A new approach to robust observer design," *Int. J. Control,* vol.47, 745–751, 1988.

103. Tsui, C., "On the solution to matrix equation $TA - FT = LC$ and its applications," *SIAM J. Matrix Anal. Appl.,* vol.14, 33–44, 1993.

104. Tzafestas, S., A. Kanellakis, and N. Theodorou, "Application of frequency dependent Lyapunov equation to 2–dimensional problems, *Proc. IEE, Part D.,* vol.139, 197–203, 1992.

105. Wachspress, E., "Iterative solution of the Lyapunov matrix equation," *Appl. Math. Letters*, vol.1, 87–90, 1988.

106. Wedderburn, J., "Note on the linear matrix equation," *Proc. Edinburgh Math. Soc.* vol.22, 49–53, 1904.

107. Wimmer, H., "Consistency of a pair of generalized Sylvester equations," *IEEE Trans. Automatic Control*, vol.39, 1014–1016, 1994.

108. Yedavalli, R., "Improved measures of stability robustness of linear state space models," *IEEE Trans. Automatic Control*, vol.30, 577-579, 1985.

109. Zhou, K. and P. Khargonekar, "Stability robustness for linear state space models with structured uncertainty," *IEEE Trans. Automatic Control*, vol.32, 621-623, 1987.

Appendix

Matrix Inequalities

In this appendix we give a summary of several matrix inequalities and one scalar inequality appearing throughout of this book.

1. Fan-Hoffman's Inequality: (Fan and Hoffman, 1955).
For any square $n \times n$ matrix Z the following holds

$$\lambda_i \left(\frac{Z + Z^T}{2} \right) \le \sigma_i(Z), \quad 1 \le i \le n$$

where λ_i and σ_i represent, respectively, the eigenvalues and singular values.

2. Fan's Singular Values Inequalities: (Fan, 1951).
For any two square $n \times n$ matrices X, Y the singular values satisfy

$$i) \quad \sigma_{j+k-1}(XY) \le \sigma_j(X)\sigma_k(Y), \quad j, k \ge 1, \ j + k \le n + 1$$
$$ii) \quad \sigma_{j+k-1}(X + Y) \le \sigma_j(X) + \sigma_k(Y), \quad j, k \ge 1, \ j + k \le n + 1$$

3. Horn's Inequality: (Horn, 1950).
For any two real symmetric positive semi-definite matrices we have

$$\prod_{i=1}^{k} \lambda_i(XY) \le \prod_{i=1}^{k} \lambda_i(X)\lambda_i(Y), \quad i = 1, 2, ..., n$$

with equality for $k = n$. The second part of Horn's inequality is given by

$$\prod_{i=1}^{k} \lambda_{n-i+1}(XY) \geq \prod_{i=1}^{k} \lambda_{n-i+1}(X)\lambda_{n-i+1}(Y), \quad k = 1, 2, ..., n$$

4. Fan's Eigenvalue Summation Inequality: (Fan, 1949).

For any $n \times n$ positive semi-definite matrices X and Y the eigenvalues satisfy

$$\sum_{i=1}^{k} \lambda_i(X + Y) \leq \sum_{i=1}^{k} \lambda_i(X) + \sum_{i=1}^{k} \lambda_i(Y), \quad k = 1, 2, ..., n$$

with equality for $k = n$. This inequality directly implies another summation inequality

$$\sum_{i=1}^{k} \lambda_{n-i+1}(X + Y) \geq \sum_{i=1}^{k} \lambda_{n-i+1}(X) + \sum_{i=1}^{k} \lambda_{n-i+1}(Y), \ k = 1, 2, ..., n$$

Note that these inequalities are not explicitly given in (Fan, 1949), but they can derived from his results (Garloff, 1986; Hmamed, 1990).

5. Fan's Eigenvalues Product Inequality: (Fan, 1953).

For an $n \times n$ matrix X, representing a linear transformation in an unitary space, if $\lambda_i(X^T + X) \geq 0$ then

$$\prod_{i=1}^{k} \lambda_{n-i+1}(X^T + X) \leq \prod_{i=1}^{k} 2Re\{\lambda_{n-i+1}(X)\}, \quad k = 1, 2, ..., n$$

6. Trace of the Product Inequality:

The following inequality can be found in (Marcus and Minc, 1965; Kleinman and Athans, 1968). For any two positive semi-definite matrices $X = X^T \geq 0$ and $Y = Y^T \geq 0$ we have

$$\lambda_{min}(X)tr(Y) \leq tr(XY) \leq \lambda_{max}(X)tr(Y)$$

In (Wang et al., 1986) is shown that the above pair of trace inequalities is also valid for $Y = Y^T \geq 0$ and $X = X^T$.

A stronger upper bound for the trace of the matrix product is obtained in (Sanjuk and Rhodes, 1987) as

$$tr(XY) \leq |tr(XY)| \leq \|X\|_2 tr(Y)$$

which is valid for $Y = Y^T \geq 0$ and X any square matrix. $\|X\|_2$ denotes the spectral norm (the largest singular value).

New stronger results for the trace of product inequalities are obtained in (Mori, 1988; Hmamed, 1989) by using the notion of the matrix measure. For a square matrix X the matrix measure is defined as

$$\mu(X) = 0.5\lambda_{max}(X + X^T)$$

Independently, (Mori, 1988) and (Hmamed, 1989) have shown that for $Y = Y^T \geq 0$ and any square matrix X one has

$$-\mu(-X)tr(Y) \leq tr(XY) \leq \mu(X)tr(Y)$$

Since

$$-\|X\|_2 \leq -\mu(-X) \leq \mu(X) \leq \|X\|_2$$

the improvement is obvious. These results are further generalized in (Fang et al., 1994) to include symmetric but not necessarily positive semi-definite matrices $Y = Y^T$ and arbitrarily square X. In addition, it has been also shown that a tighter upper bound can be obtained for $Y > 0$ than the one of (Mori, 1988; Hmamed, 1989).

7. Mirsky Inequality: (Mirsky, 1959).
For any two real symmetric matrices the following trace inequalities hold

$$\sum_{i=1}^{n} \lambda_i(X)\lambda_{n-i+1}(Y) \leq tr(XY) \leq \sum_{i=1}^{n} \lambda_i(X)\lambda_i(Y)$$

8. Ostrowski Inequalities:
Ostrowski inequalities (Ostrowski, 1959) for the eigenvalues of a sum of symmetric matrices and for the eigenvalues of a matrix product.

8.1) *Matrix Sum*: Let X and Y be symmetric matrices of dimension n. Then the eigenvalues satisfy

$$\lambda_i(X + Y) = \lambda_i(X) + h_i, \qquad i = 1, 2, ..., n$$

where

$$\lambda_1(Y) \geq h_i \geq \lambda_n(Y)$$

or by using notation from Theorem 3.1

$$\lambda_i(X) + \lambda_1(Y) \geq \lambda_i(X + Y) \geq \lambda_i(X) + \lambda_n(Y)$$

8.2) *Matrix Product*: Let $P = P^T$ and A be square matrices, then the eigenvalues satisfy

$$\lambda_i(A^T P A) = \lambda_i(A A^T P) = \theta_i \lambda_i(P), \quad \sigma_1^{1/2} \geq \theta_i \geq \sigma_n^{1/2}$$

9. Weyl Inequality:

Let X and Y be hermitian matrices of dimension n. Then, the eigenvalues of a matrix sum satisfy

$$\lambda_{i+j-n}(X + Y) \geq \lambda_i(X) + \lambda_j(Y), \ \ i, j = 1, 2, ..., n$$
$$i + j \geq n + 1$$

$$\lambda_{i+j-1}(X + Y) \leq \lambda_i(X) + \lambda_j(Y), \ \ i, j = 1, 2, ..., n$$
$$i + j \leq n + 1$$

Weyl inequality can be found in (Amir-Moez, 1956).

10. Amir-Moez Inequality: (Amir-Moez, 1956).

For any $n \times n$ positive semi-definite matrices X and Y the following eigenvalue inequalities hold

$$\lambda_{i+j-1}(XY) \leq \lambda_i(X)\lambda_j(Y), \quad if \ \ i + j \leq n + 1$$
$$\lambda_{i+j-n}(XY) \geq \lambda_i(X)\lambda_j(Y), \quad if \ \ i + j \geq n + 1$$

11. Coppel's Inequality: (Coppel, 1965).

For any real square $n \times n$ matrix X the following is valid

$$\lambda_{max}\left(e^{Xt}e^{X^Tt}\right) \leq e^{2\mu(X)t}, \quad \mu = \frac{1}{2}\lambda_1\left(X + X^T\right), \quad t \geq 0$$

Another version of Coppel's inequality is derived in (Mori et al., 1987)

$$\lambda_{min}\left(e^{Xt}e^{X^Tt}\right) \geq e^{-2\mu(-X)t}, \quad \mu = \frac{1}{2}\lambda_1\left(X + X^T\right), \quad t \geq 0$$

12. Minkovski Inequality: (Marcus and Minc, 1964).

For any two square matrices the following property of determinant holds

$$|X + Y|^{1/n} \geq |X|^{1/n} + |Y|^{1/n}$$

13. Inequality A: (Beckenbach and Bellman, 1965).

Let $Y = Y^T > 0$, $X = X^T > 0$, and $Y, X \in \Re^{n \times n}$, then determinants satisfy

$$|X|^{1/n} = \frac{min}{|Y|}\left[\frac{1}{n}tr\{XY\}\right]$$

14. Inequality B: (Patel and Toda, 1979).

Let $Y > 0$ and $Y, X \in \Re^{n \times n}$, then

$$tr\{Y^{-1}XYX^T\} \geq \sum_{i=1}^{n} |\lambda_i(X)|^2 \geq \frac{1}{n}[tr\{X\}]^2$$

15. Inequality C: The following matrix inequality can be found in (Mori, 1986)

$$\lambda_{min}(X)YY^T \leq YXY^T \leq \lambda_{max}(X)YY^T, \quad X = X^T$$

16. Scalar Arithmetic-Mean Geometric-Mean Inequality:

$$\left(\prod_{i=1}^{n} x_i\right)^{1/n} \leq \frac{1}{n}\sum_{i=1}^{n} x_i, \quad x_i \geq 0$$

This very well-known inequality can be found in many books, for example (Mitrinovic, 1970).

Several additional useful matrix inequalities are given in (Patel and Toda, 1979; Garloff, 1986; Komaroff, 1989). Some scalar inequalities encountered in the study of bounds of the solution attributes of the Lyapunov equation can be found in (Komaroff, 1992).

References

1. Amir-Moez, A., "Extreme properties of a Hermitian transformation and singular values of the sum and product of linear transformation," *Duke Math., J.*, vol.23, 463–476, 1956.
2. Bechenbach, E. and R. Bellman, *Inequalities*, Springer Verlag, Berlin, 1965.
3. Coppel, W., *Stability and Asymptotic Behavior of Differential Equations,* Heath and Company, Boston, 1965.
4. Fan, K., "On the theorem of Weyl concerning eigenvalues of linear transformations," *Proc. National Academy of Science, USA*, vol.35, 652–655, 1949.
5. Fan, K., "Minimum properties and inequalities for the eigenvalues of completely continuous operators," *Proc. National Academy of Science, USA*, vol.37, 760–766, 1951.
6. Fan, K., "A minimum property of the eigenvalues of a Hermitian transformation," *Amer. Math. Monthly*, vol.60, 48–50, 1953.
7. Fan, K. and A. Hoffman, "Some metric inequalities in the space of matrices," *Proc. Amer. Math. Soc.*, vol.6, 111–116, 1955.
8. Fang, Y., K. Loparo, and X. Feng, "Inequalities for the trace of matrix product," *IEEE Trans. Automatic Control*, vol.39, 2489–2490, 1994.
9. Garloff, J., "Bounds for eigenvalues of the solution of the discrete Riccati and Lyapunov equations and the continuous Lyapunov equation," *Int. J. Control*, vol.43, 423-431, 1986.
10. Hmamed, A., "A matrix inequality," *Int. J. Control*, vol.49, 363-365, 1989.
11. Horn, A., "On the singular values of a product of completely continuous operators," *Proc. National Academy of Science, USA*, vol.36, 374–375, 1950.
12. Kleinman, D. and M. Athans, "The design of suboptimal time-varying systems," *IEEE Trans. Automatic Control*, vol.13, 150–159, 1968.

13. Komaroff, N., "Matrix inequalities applicable to estimating solution sizes of Riccati and Lyapunov equations," *IEEE Trans. Automatic Control*, vol.34, 97–98, 1989.

14. Komaroff, N., "Upper summation and product bounds for solution eigenvalues of the Lyapunov matrix equation," *IEEE Trans. Automatic Control*, vol.37, 1040–1042, 1992.

15. Marcus, M. and H. Minc, *A Survey of Matrix Theory and Inequalities*, Allyn and Bacon, Boston, 1964.

16. Mirsky, L., "On the trace of matrix product," *Math. Nachr.*, vol.20, 171–174, 1959.

17. Mitrinovic, D., *Analytic Inequalities*, Springer Verlag, New York, 1970.

18. Mori, T., N. Fukuma, and M. Kuwahara, "Explicit solution and eigenvalue bounds in the Lyapunov matrix equation," *IEEE Trans. Automatic Control*, vol.31, 656-658, 1986.

19. Mori, T., "Comments on 'A matrix inequality associated with bounds on solutions of algebraic Riccati and Lyapunov equations'," *IEEE Trans. Automatic Control*, vol.33, 1088, 1988.

20. Ostrowski, A., "A quantitative formulation of Sylvester's law of inertia," *Proc. National Academy of Science, USA*, vol.45, 740–744, 1959.

21. Patel, R. and M. Toda, "Trace inequalities involving hermitian matrices," *Linear Algebra and Its Appl.*, vol.23, 13–20, 1979.

22. Sanjuk, J. and I. Rhodes, "A matrix inequality associated with bounds on solutions of algebraic Riccati and Lyapunov equations," *IEEE Trans. Automatic Control*, vol.32, 739, 1987.

23. Wang, S., T. Kuo, and H. Hsu, "Trace bounds on the solution of the algebraic matrix Riccati and Lyapunov equations," *IEEE Trans. Automatic Control*, vol.31, 654–656, 1986.

Index

Engineering

DE RE METALLICA, Georgius Agricola. The famous Hoover translation of greatest treatise on technological chemistry, engineering, geology, mining of early modern times (1556). All 289 original woodcuts. 638pp. 6¾ x 11. 0-486-60006-8

FUNDAMENTALS OF ASTRODYNAMICS, Roger Bate et al. Modern approach developed by U.S. Air Force Academy. Designed as a first course. Problems, exercises. Numerous illustrations. 455pp. 5⅜ x 8½. 0-486-60061-0

DYNAMICS OF FLUIDS IN POROUS MEDIA, Jacob Bear. For advanced students of ground water hydrology, soil mechanics and physics, drainage and irrigation engineering and more. 335 illustrations. Exercises, with answers. 784pp. 6⅛ x 9¼.
0-486-65675-6

THEORY OF VISCOELASTICITY (Second Edition), Richard M. Christensen. Complete consistent description of the linear theory of the viscoelastic behavior of materials. Problem-solving techniques discussed. 1982 edition. 29 figures. xiv+364pp. 6⅛ x 9¼. 0-486-42880-X

MECHANICS, J. P. Den Hartog. A classic introductory text or refresher. Hundreds of applications and design problems illuminate fundamentals of trusses, loaded beams and cables, etc. 334 answered problems. 462pp. 5⅜ x 8½. 0-486-60754-2

MECHANICAL VIBRATIONS, J. P. Den Hartog. Classic textbook offers lucid explanations and illustrative models, applying theories of vibrations to a variety of practical industrial engineering problems. Numerous figures. 233 problems, solutions. Appendix. Index. Preface. 436pp. 5⅜ x 8½. 0-486-64785-4

STRENGTH OF MATERIALS, J. P. Den Hartog. Full, clear treatment of basic material (tension, torsion, bending, etc.) plus advanced material on engineering methods, applications. 350 answered problems. 323pp. 5⅜ x 8½. 0-486-60755-0

A HISTORY OF MECHANICS, René Dugas. Monumental study of mechanical principles from antiquity to quantum mechanics. Contributions of ancient Greeks, Galileo, Leonardo, Kepler, Lagrange, many others. 671pp. 5⅜ x 8½. 0-486-65632-2

STABILITY THEORY AND ITS APPLICATIONS TO STRUCTURAL MECHANICS, Clive L. Dym. Self-contained text focuses on Koiter postbuckling analyses, with mathematical notions of stability of motion. Basing minimum energy principles for static stability upon dynamic concepts of stability of motion, it develops asymptotic buckling and postbuckling analyses from potential energy considerations, with applications to columns, plates, and arches. 1974 ed. 208pp. 5⅜ x 8½.
0-486-42541-X

METAL FATIGUE, N. E. Frost, K. J. Marsh, and L. P. Pook. Definitive, clearly written, and well-illustrated volume addresses all aspects of the subject, from the historical development of understanding metal fatigue to vital concepts of the cyclic stress that causes a crack to grow. Includes 7 appendixes. 544pp. 5⅜ x 8½. 0-486-40927-9

Mathematics

FUNCTIONAL ANALYSIS (Second Corrected Edition), George Bachman and Lawrence Narici. Excellent treatment of subject geared toward students with background in linear algebra, advanced calculus, physics and engineering. Text covers introduction to inner-product spaces, normed, metric spaces, and topological spaces; complete orthonormal sets, the Hahn-Banach Theorem and its consequences, and many other related subjects. 1966 ed. 544pp. 6⅛ x 9¼. 0-486-40251-7

ASYMPTOTIC EXPANSIONS OF INTEGRALS, Norman Bleistein & Richard A. Handelsman. Best introduction to important field with applications in a variety of scientific disciplines. New preface. Problems. Diagrams. Tables. Bibliography. Index. 448pp. 5⅜ x 8½. 0-486-65082-0

VECTOR AND TENSOR ANALYSIS WITH APPLICATIONS, A. I. Borisenko and I. E. Tarapov. Concise introduction. Worked-out problems, solutions, exercises. 257pp. 5⅝ x 8¼. 0-486-63833-2

AN INTRODUCTION TO ORDINARY DIFFERENTIAL EQUATIONS, Earl A. Coddington. A thorough and systematic first course in elementary differential equations for undergraduates in mathematics and science, with many exercises and problems (with answers). Index. 304pp. 5⅜ x 8½. 0-486-65942-9

FOURIER SERIES AND ORTHOGONAL FUNCTIONS, Harry F. Davis. An incisive text combining theory and practical example to introduce Fourier series, orthogonal functions and applications of the Fourier method to boundary-value problems. 570 exercises. Answers and notes. 416pp. 5⅜ x 8½. 0-486-65973-9

COMPUTABILITY AND UNSOLVABILITY, Martin Davis. Classic graduate-level introduction to theory of computability, usually referred to as theory of recurrent functions. New preface and appendix. 288pp. 5⅜ x 8½. 0-486-61471-9

ASYMPTOTIC METHODS IN ANALYSIS, N. G. de Bruijn. An inexpensive, comprehensive guide to asymptotic methods–the pioneering work that teaches by explaining worked examples in detail. Index. 224pp. 5⅜ x 8½ 0-486-64221-6

APPLIED COMPLEX VARIABLES, John W. Dettman. Step-by-step coverage of fundamentals of analytic function theory–plus lucid exposition of five important applications: Potential Theory; Ordinary Differential Equations; Fourier Transforms; Laplace Transforms; Asymptotic Expansions. 66 figures. Exercises at chapter ends. 512pp. 5⅜ x 8½. 0-486-64670-X

INTRODUCTION TO LINEAR ALGEBRA AND DIFFERENTIAL EQUATIONS, John W. Dettman. Excellent text covers complex numbers, determinants, orthonormal bases, Laplace transforms, much more. Exercises with solutions. Undergraduate level. 416pp. 5⅜ x 8½. 0-486-65191-6

RIEMANN'S ZETA FUNCTION, H. M. Edwards. Superb, high-level study of landmark 1859 publication entitled "On the Number of Primes Less Than a Given Magnitude" traces developments in mathematical theory that it inspired. xiv+315pp. 5⅜ x 8½. 0-486-41740-9

TENSOR CALCULUS, J.L. Synge and A. Schild. Widely used introductory text covers spaces and tensors, basic operations in Riemannian space, non-Riemannian spaces, etc. 324pp. 5⅜ x 8¼. 0-486-63612-7

ORDINARY DIFFERENTIAL EQUATIONS, Morris Tenenbaum and Harry Pollard. Exhaustive survey of ordinary differential equations for undergraduates in mathematics, engineering, science. Thorough analysis of theorems. Diagrams. Bibliography. Index. 818pp. 5⅜ x 8½. 0-486-64940-7

INTEGRAL EQUATIONS, F. G. Tricomi. Authoritative, well-written treatment of extremely useful mathematical tool with wide applications. Volterra Equations, Fredholm Equations, much more. Advanced undergraduate to graduate level. Exercises. Bibliography. 238pp. 5⅜ x 8½. 0-486-64828-1

FOURIER SERIES, Georgi P. Tolstov. Translated by Richard A. Silverman. A valuable addition to the literature on the subject, moving clearly from subject to subject and theorem to theorem. 107 problems, answers. 336pp. 5⅜ x 8½. 0-486-63317-9

INTRODUCTION TO MATHEMATICAL THINKING, Friedrich Waismann. Examinations of arithmetic, geometry, and theory of integers; rational and natural numbers; complete induction; limit and point of accumulation; remarkable curves; complex and hypercomplex numbers, more. 1959 ed. 27 figures. xii+260pp. 5⅜ x 8½. 0-486-63317-9

POPULAR LECTURES ON MATHEMATICAL LOGIC, Hao Wang. Noted logician's lucid treatment of historical developments, set theory, model theory, recursion theory and constructivism, proof theory, more. 3 appendixes. Bibliography. 1981 edition. ix + 283pp. 5⅜ x 8½. 0-486-67632-3

CALCULUS OF VARIATIONS, Robert Weinstock. Basic introduction covering isoperimetric problems, theory of elasticity, quantum mechanics, electrostatics, etc. Exercises throughout. 326pp. 5⅜ x 8½. 0-486-63069-2

THE CONTINUUM: A CRITICAL EXAMINATION OF THE FOUNDATION OF ANALYSIS, Hermann Weyl. Classic of 20th-century foundational research deals with the conceptual problem posed by the continuum. 156pp. 5⅜ x 8½. 0-486-67982-9

CHALLENGING MATHEMATICAL PROBLEMS WITH ELEMENTARY SOLUTIONS, A. M. Yaglom and I. M. Yaglom. Over 170 challenging problems on probability theory, combinatorial analysis, points and lines, topology, convex polygons, many other topics. Solutions. Total of 445pp. 5⅜ x 8½. Two-vol. set. Vol. I: 0-486-65536-9 Vol. II: 0-486-65537-7

Paperbound unless otherwise indicated. Available at your book dealer, online at **www.doverpublications.com**, or by writing to Dept. GI, Dover Publications, Inc., 31 East 2nd Street, Mineola, NY 11501. For current price information or for free catalogues (please indicate field of interest), write to Dover Publications or log on to **www.doverpublications.com** and see every Dover book in print. Dover publishes more than 500 books each year on science, elementary and advanced mathematics, biology, music, art, literary history, social sciences, and other areas.